THE OCEAN RANGER

REMAKING THE PROMISE OF OIL

Susan Dodd

FERNWOOD PUBLISHING
HALIFAX & WINNIPEG

Editing: Brenda Conroy
Cover Design: John van der Woude
Cover Photo: Kazuko Kizawa, 2011
Printed and bound in Canada by Hignell Book Printing

Published in Canada by Fernwood Publishing
32 Oceanvista Lane
Black Point, Nova Scotia, B0J 1B0
and 748 Broadway Avenue, Winnipeg, Manitoba, R3G 0X3
www.fernwoodpublishing.ca

Fernwood Publishing Company Limited gratefully acknowledges the financial support of the Government of Canada through the Canada Book Fund, the Canada Council for the Arts, the Nova Scotia Department of Tourism and Culture, the Manitoba Department of Culture, Heritage and Tourism under the Manitoba Publishers Marketing Assistance Program and the Province of Manitoba, through the Book Publishing Tax Credit, for our publishing program.

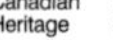

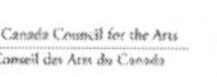

Library and Archives Canada Cataloguing in Publication

Dodd, Susan, 1966-
The Ocean Ranger : remaking the promise of oil / Susan Dodd.

Includes bibliographical references.
ISBN 978-1-55266-464-3

1. Ocean Ranger (Drilling rig) 2. Offshore oil well drilling--Accidents--Social aspects--Newfoundland and Labrador. 3. Offshore oil well drilling--Accidents--Political aspects--Newfoundland and Labrador. I. Title.

TN871.3.D63 2011 363.11'96223381909718 C2011-907950-X

CONTENTS

FOR JOYCE AND ED

I'LL SELL THEM TO THEIR ENEMIES.

—DYLAN, "WORKING MAN BLUES NO. 2"

ACKNOWLEDGEMENTS

Thank you to everyone who supported me through this project. The collegial community of the University of King's College was a great help. Thanks especially to the summer writing groups at the college, organized by Dr Melanie Frappier. Dr Elizabeth Edwards, Dr Leticia Meynell, Dr Melanie Frappier, Dr Kathryn Morris and Prof Sue Newhook all read and commented on early drafts of chapters. Lisa Mullins gave me exacting feedback on early versions. The University of King's College Research and Travel Funds supported me in three research trips and to present an early version of "Blood Money" at the American Law and Society Association annual conference in Chicago in June 2010. Funds from the Social Sciences and Humanities Research Council enabled me to hire Colin Rose, who developed a bibliography for "Blood Money" through the King's College Aid to Small Universities Grant. Delaine Tiniakos-Doran, my research assistant through the King's Student Assistant Programme, found key court rulings. Thank you to my students in the *Foundation Year Programme* for pressing the question of history's relation to justice.

Lisa Moore, Mike Heffernan, Dr Douglas House, Dr Sue Hart and Ray Hawco were all generous as authors who have written on the *Ocean Ranger* disaster.

Thanks to the following people who gave me interviews and/or provided ideas: T. Alex Hickman, Dr Jerome Davis, Benton Musselwhite Sr, Ray Wagner, Ches Crosbie, Douglas Moores, Marie Wadden, Glenn Carter, Cle Newhook, Lorraine Michael MHA and Leader NLNDP, Dr Sue Hart, Howard Pike, Prof Martin Davies (Tulane), Frank Spangnoletti, Jack Harris MP, John Crosbie, Gord Graham.

Thank you to Michael V. Coyle J.D., Catherine Campbell, Stanley Campbell, Dr Matthew Furlong, Dr Dorota Glowacka, Dr Neil Robertson and Brian O'Neill for reading parts of drafts.

Thank you to librarians, especially at the Centre for Newfoundland Studies at Memorial University, the Legislative Library at the Newfoundland and Labrador House of Assembly, Dalhousie Law School and the Public Archives in St John's. Thanks also to the instructors at the Marine Institute for the tour of their facility and demonstration of a training session.

I owe a special thank you to family members of men lost on the *Ocean Ranger* who agreed to speak with me: Scotty Morrison, Pat Ryan, Cynthia

Parsons-Walker, Samantha Gerbeau, Noreen O'Neill, Margaret Blackmore and Brian Bursey. These family members encouraged me all through this project. To family members I have not had a chance to meet, I hope that my effort at telling part of this story is helpful to you even in some small way.

As for my parents, Joyce and Ed Dodd, at the end of my most recent visit with them my father helped me load my car and then turned from the veranda of the house in Berwick and called out, "Don't let them make you water it down!" Each of them has supported me immeasurably in this project, as in everything else.

Thank you as well to Chris Jones for speaking with me so soon following the loss of his brother, Gordon Jones, on the *Deepwater Horizon*.

I am forever indebted to Dr Laura Penny, Mary Campbell, Ed Dodd and Dr Melanie Frappier, who generously read and commented on full drafts of this book.

Thank you to Errol Sharpe of Fernwood Publishing and to the production and promotion people: Brenda Conroy, Beverley Rach, Debbie Mathers and Nancy Malek.

All omissions and errors are my own.

Preface

FEBRUARY 14–15, 1982

"They've gone to the lifeboats and we're waiting for news."

Searching for oil under the Atlantic Ocean, the *Ocean Ranger* and its eighty-four-man crew were anchored in the Hibernia oilfield off the east coast of the Canadian province of Newfoundland on February 14, 1982. It was the heart of winter, in the early days of Canada's offshore drilling.

Designed, owned and operated by the New Orleans-based oil giant, the Ocean Drilling and Exploration Company (ODECO), the rig was under contract to another American petroleum mammoth, Mobil Oil, which managed the drilling of the well. Built by Mitsubishi and staffed by Schlumberger and others, the *Ocean Ranger* was the largest semi-submersible oil rig in the world. Deemed "unsinkable," the rig was a symbol of cutting-edge marine oil exploration technology.

But that Valentine's night, the rig was hit by an awe-inspiring North Atlantic storm. Waves crashed against the rig's giant legs, one of which housed a room where men monitored and controlled the rig's stability by letting water in and out of its massive pontoons. A portlight in the ballast room was smashed during the rising storm, and salt water rushed in to soak the ballast control panel. Its lights flashed erratically.

No one knew what to do.

The men had no formal training. They had no marine captain who understood the rig. There was no one ashore they trusted to advise them. Their operating manual was incomplete at best, and their training had been misleading at worst. If the men had closed the portlight, mopped up the water and gone to sleep, they might have lived to see the dawn. As it was, the flashing lights on the ballast control panel falsely indicated that water was moving randomly in and out of the pontoons. The men cut the power to the panel. The rig tilted slightly. The men intervened manually and then turned the power back on. The rig's slight tilt became a dangerous list. Waves crashed into the wells that held the anchor chains and pulled the rig further off balance. While the February storm raged, the men called in their mayday and disappeared from the airwaves.

It is very hard to think about that evacuation.

Much of the *Ocean Ranger*'s safety equipment did not work. The men were not trained to muster at designated boats or to launch and operate the lifeboats. They did not have cold water immersion suits. Some of the men—we will never know who—jumped into the February North Atlantic in their jeans and flannel shirts.

The rig's supply ship, the *Seaforth Highlander*, steamed towards the *Ranger*'s last known position. They were an hour away because they had no established protocol to stay close in bad weather. When the *Seaforth Highlander* finally arrived, its crew found a boat full of living men who had somehow escaped being crushed against the sinking rig by the wind and waves.

Not knowing, as Mobil head office knew, that a rope attached to that kind of lifeboat in rough seas will cause it to capsize, the would-be rescuers on the *Highlander* secured lines to it. The men in the lifeboat crowded out of the enclosed capsule and the boat flipped.

The men plunged into the sea. One man from the *Ocean Ranger*'s crew nearly reached the hand of a crewman on the *Seaforth Highlander*, but the storm pulled him away. Despite the ship's role as an emergency resource for the rig, the *Highlander*'s crew had no equipment designed to pluck men from the sea.

• • •

The telephone cut through the early morning at our house in Berwick, Nova Scotia and I heard my mother run to answer the call.

"You might as well tell her," Mom said hours later when I joined my parents in the kitchen.

My eldest brother had a job on the *Ocean Ranger*.

"Jim's rig is in trouble. They've gone to the lifeboats and we're waiting for news," said my Dad, in his best "Major Dodd" voice. My father was retired from the airforce, and he had a good idea of what it must be like out there that night because he had flown many search and rescue missions over the wintry North Atlantic.

Jim was the oldest, most adventurous and biggest trouble-maker of us five Dodd children. As my mother recalls, "None of you are stupid, but Jim… Jim was really something…" I remember Jim, defiant after his arrest for selling hashish at the family cottage. With his hair to his shoulders, he split wood behind the house, waiting to serve two months in jail for a charge that today would get him no more than a small fine: "Paying my debt to society," he explained sarcastically. When we talked, Jim liked to mock aged wisdom addressing naive youth. "Suzie, the only two worse things I could have done for Dad's career would be to be a homosexual or a communist. A drug-dealing homo-commie would have it all," he joked. Jim loved reading and music, and though I was

Capsized Ocean Ranger *lifeboat.*
(Courtesy of Canadian Armed Forces and Canadian Press.)

still in elementary school, he gave me his university English books, including *Equus*, "because you like horses." My bookish ways amused him. He also gave me the Stones' *Exile on Main Street*—we grew up on Main Street.

I was ten and he was eighteen when he went to jail for those two months. Two months is a long time to spend in jail when you only live to be twenty-four. It seemed long that summer and it seems even longer now, considering that the people responsible for Jim's death were never charged or fined, let alone locked up. They're out there still, running oil rigs, or they're retired, somewhere warm.

Jim hitch-hiked out West when I was in grade four. He started in the oil fields with all the other "Eastern creeps and bums," as Ralph Klein, then mayor of Calgary, called them. They went out to make a little money and have an adventure before returning home. Jim came home every summer. He did well out West, studying at community college to become a "mud engineer," analyzing the ooze used in drilling. One of the lucky ones, Jim got a job back on the East Coast. The pay and work schedule were so great that he didn't even mind the required "corporate disco hair cut," as he called it. He was talking about continuing university and getting along with "the folks" for the first time in ages.

The night before he flew out for the last time, he stayed up late in the warm kitchen with Mom, eating cookies and talking.

We hoped during the search, but unreasonably, and we knew it. Between phone calls from Jim's distraught shore manager, Mom and Dad played cards. My brothers Ian and Ron came home from Acadia University, in nearby Wolfville, and my brother Dave flew home from a surveying job in northern Alberta. Casseroles arrived. CBC news anchor Knowlton Nash pronounced all eighty-four men "missing and presumed dead," though they did recover twenty-two bodies. And that capsized lifeboat heaved and heaved in the grey television waves

They didn't find Jim's body, so we had our memorial service and wake without him. "Mobil Oil will pay for the beer." We laughed.

Introduction

THE PROMISE OF OIL... BROKEN AND REMADE

"The government had us convinced that the oil industry was... the start of a new era. We had this... unsinkable rig, but then it sank and all these people died. In a cultural sense, so did our dream."

My brother Jim's death was a personal loss for me and my family, but it was also a political failure. It would take me decades to realize the wider importance of the *Ocean Ranger* disaster and to marvel at the effectiveness of our political and economic systems' ability to manage shocks to public confidence. The loss of the *Ocean Ranger* and its crew on February 15, 1982, betrayed the promise that offshore oil would liberate Newfoundland from economic dependency on central Canada and open the way for cultural rebirth and self-determination. This "promise of oil" was reconstituted over the course of the thirty years that followed the disaster, to the point where Danny Williams could retire in triumph as Premier of Newfoundland and Labrador, declaring: "We fought the federal government for equity under the Atlantic Accord, and we won. We fought the big oil companies for more benefits, and we won. And today, we actually share a tremendous partnership and a mutual respect with them" (2010). In Premier Williams' account, the Newfoundland people now knew themselves as masters of their collective destiny, wresting benefits and recognition from grudging outsiders. "Our time has come. Our future is being shaped in our own hands. We are the master sculptor holding in our hands the clay that will mould and form our destiny."

For Newfoundland to be poised like this, ready to shape its own future, the past needed to be mastered, especially the traumatic past. The loss of those eighty-four men and the supposedly state-of-the-art *Ocean Ranger* inflicted an injury on the public of Newfoundland. So shaken was the belief in the promise of oil that in 1983, though Newfoundland had no legal right to do so, the province banned offshore drilling. The federal government countered that order, and the Ocean Ranger Families Foundation accused politicians of using the emotions stirred by the anniversary of the disaster strategically. A shock in its own right, the loss of the crew and the rig also resonated with older, foundational injuries suffered by "the people of Newfoundland." The *Ocean Ranger* disaster threatened to revive colonial fears of inferiority. It menaced a genera-

Leo Barry, Newfoundland Energy Minister, and the vial of oil, 1979 (Courtesy of Canadian Press)

tion of ambitious men in the province—and in Ottawa—who felt they were poised to solve the riddle of Newfoundland's resource-rich but economically poor history. The solution to the riddle was simple: local control over offshore oil development.

This book considers how the "promise of oil" was remade in the wake of the traumatic loss of the *Ocean Ranger* and its eighty-four crew members. Powerful men in Ottawa, St John's, New York, Houston and New Orleans surely asked: "How can we make this go away?" Such self-conscious attempts to control the story had to grapple with the workings of unconscious recollections at personal and collective levels. The socio-political processes by which the disastrous corporate and government neglect at the root of the *Ocean Ranger* loss was rewritten as a story of collective striving and cultural maturation are complex. The *Ocean Ranger* story took shape in historical memory while Newfoundland

The Ocean Ranger (Publicity photo from ODECO, *courtesy Mike Heffernan)*

and Canada staved off a potential crisis in public confidence in the promise of oil. In the first half of this book, I detail the ways our political and economic systems shored up public acceptance of their legitimacy as protectors of the very people they had allowed to be killed and bereaved in the disaster. Out of this description of the aftermath of "my" disaster, I develop a way of analyzing the aftermaths of industrial disasters and collective traumas more broadly considered, especially when they result from the failure of governments to regulate corporations.

The shock of the *Ocean Ranger* disaster was not that oil production was dangerous, but rather the realization that governments had betrayed people's faith. People trusted governments to use reasonable regulation to mitigate the risks of oil jobs. That trust was misplaced. There were no provincial safety regulations in the Newfoundland offshore when my brother and his eighty-three co-workers died (O'Neill 1988: 155). Time and again, publics trust governments to ensure that companies operate with reasonable prudence. Time and again we are shocked by a new disaster caused by corporate negligence. We say we will "never forget." Then we forget. And then it happens again. The most recent example being in 2010 when the *Deepwater Horizon* disaster killed eleven workers, injured seventeen more and resulted in the worst U.S. marine oil spill in history. Like the *Ocean Ranger* disaster the *Deepwater Horizon* disaster was caused, at its root, by what President Obama referred to as "the cosy relationship between the oil companies and the federal agency that permits them to drill" (BBC 2010). BP and Transocean's "cozy relationship" with U.S. regulators made a mockery of environmental and safety regulations for deepwater drilling in the Gulf of Mexico. The *Ocean Ranger* workers' testimony should have been ringing in our ears in 2011 when Transocean's Doug Brown testified that on the *Deepwater Horizon* drilling continued despite safety protocols that should have interrupted that tragic "chain of events."

THE PROMISE OF OIL, CIRCA 1982

In 1979, when Chevron struck oil in the Hibernia field, politicians realized the extent of the petroleum deposits on the Grand Banks. Newfoundland Energy Minister Leo Barry famously posed in front of a map of the province holding up a jar of oil. The Newfoundland and Canadian governments engaged in a lengthy and noisy battle for jurisdiction over the offshore, reopening lingering grievances from the bitter 1948 referendum, which saw only 52 percent of Newfoundlanders choose to join Confederation. The dispute over control of offshore petroleum development struck all the chords of nationalist sentiment: the physical boundary of the province as a distinct territory, its loss of sovereignty on joining Canada and, most importantly, its view of itself as a community of individuals unified by an imagined identity rather than direct personal interaction or even shared material interests (Anderson 2003: 7; Marland 2010).

Oil money was a hot topic in Canada in the 1970s and 1980s, with the Liberal federal government trying to staunch the flow of oil profits into the United States with the National Energy Program. By the time of the *Ocean Ranger* loss, Canadians and Newfoundlanders had invested heavily to entice oil companies to Canada's offshore. In return, the oil companies offered jobs and an entry into the world's most lucrative industry. With tax credits and deductions covering the vast majority of the cost of discovering the Hibernia and Ventura oil fields,[1] one would have thought that the two levels of government could not afford to squabble in the immediate aftermath of the *Ocean Ranger* disaster. The *Ocean Ranger* capsized during the exploration phase, before the oil and money began to flow into Newfoundland and beyond. If the federal and provincial governments looked like they were so busy fighting over oil money that they left our men exposed to the February seas, there was a danger the public would lose confidence in the promise of oil (O'Neill 1988: 166; Heffernan 2009; Cadigan 2009: 269).

If Newfoundland could become a player in the international petroleum industry, the former colony could assume a new, confident, collective identity. Entry into the international oil club promised a maturity sought by Atlantic Canada generally. This is a theme that played out in the lives of so many of the men on the *Ranger*, including my brother Jim. My parents like to tell an old story about a chance encounter with Jim. They were driving across the Prairies in the late 1970s and stopped at a highway diner. There, slouched at the counter, was Jim. Polishing the cutlery with his flannel shirt tail, he advised jokingly, "You gotta watch the hygiene in these greasy spoons." Mom and Dad were in Jim's world, the Trans-Canada highway between the oil fields of Alberta and the Atlantic coast. "We met him as friends, instead of as parents," Mom remembers. They ate together, then Jim sped off in his "little foreign sports car," an old Toyota Celica with a feathered roach clip swinging from the rearview mirror. For my parents, Jim's taking the job on the *Ocean Ranger* was a sign that he was growing up: he got a haircut, he marketed the skills he had acquired at college, and he looked towards a future of study and travel, thanks to the oil industry.

Landing a job on the East coast in the early 1980s was good luck. Unemployment rates were soaring—almost 15 percent of Newfoundlanders and 10 percent of Nova Scotians were out of work between 1977 and 1981 (Statistics Canada 2011). Offshore oil offered work to the experienced and to newcomers, to the highly salaried and to wage labourers. For many of the single men, oil rig work was a starter job. Craig Tilley joined the *Ocean Ranger*'s crew shortly before the disaster. He had recently graduated from Gonzaga High School in St John's, and, at nineteen, he was starting to settle down after some restless years. Tilley was the youngest man on the rig that night. More established workers were able to move home: Paul Bursey was thirty, single and, after moving back and forth between Newfoundland and the mainland for years, he became his widowed mother's main support when he took the St John's-based job on the *Ocean Ranger*. Newfoundland's offshore was even creating jobs for ambitious young men from central Canada. Perry Morrison grew up in Toronto, and, university degree in hand, he mystified his parents by announcing: "I am going to become a deep sea diver." Newfoundland's offshore was to be Perry's gateway to a lucrative international career in underwater high-pressure welding.

The *Ocean Ranger* job was a chance for men with young wives and children to build towards the North American dream of providing their families a stable and nurturing home. Clyde Parsons was an autobody worker laid off from the garage, and he begged his way onto the rig. He and Cynthia Parsons had two children, and a cab driver I met in St John's recalled: "I worked with Clyde years ago at the shop. Even when he was really young he was a family man—did his work and went home with the kids. It wasn't the way it is now, it wasn't the fashion then. He just seemed to enjoy it." Paschal O'Neill's son Ashley was nine months old when his wife Noreen dropped him and some buddies at the heliport for that hitch on the rig: "He never talked about dangers out there, but this day they got joking in the car. 'Look, here he comes with a brief case—as if there's anything in there about how to work a rig!'"—they joked about one key ODECO employee. Noreen drove home slowly that day, thinking that if his flight were held up, she would just bring Paschal home and say "to hell with this job" (O'Neill 2010).

THE BROKEN PROMISE

By the time family members heard that the rig was in trouble, the men were lost, the rig had capsized, and Mobil and ODECO were preparing their strategic responses. Ever since I first read Chief Justice T. Alex Hickman's inquiry report in 1984, I have tried to imagine what the managers and executives of Mobil Oil and ODECO did in the mysterious hours between 3:38 a.m., when the rig disappeared from radar, and 7:35 a.m., when Mobil finally shared that information with Canadian Search and Rescue. Who was woken from his sleep? Did they delay announcing the disaster to develop public relations and legal strategies? Is that wrong, given that the men were already beyond help?

Some time after daylight, I walked to the end of Main Street to retrieve our wayward dog, Quincy, from a neighbour's, while Mom and Dad waited by the telephone. The storm was so severe that where we lived, in Nova Scotia about a thousand kilometres away from the *Ocean Ranger*, the sidewalks were shin-deep with snow. Across Newfoundland, the morning news reported that the rig was listing, and, echoing the last words from Ken Blackmore, the *Ranger*'s radio operator, reports announced that the men had "gone to the lifeboats." Friends and neighbours waded through drifts to the homes of young women with babies, parents in their early forties with sons freshly launched into the workforce and elderly widows with grown sons. Fifty-six of the dead men were Newfoundlanders, fifteen were Americans, twelve were from other Canadian provinces, and one was British. It seemed that everyone—except ODECO—was hoping to get to the families before they heard the news on the radio. One son heard the phone ring around 8:00 a.m. His mother answered. "She was told that they were having some communication problems with the *Ocean Ranger*. Communications had broken off" (Clayton Burry, in House 1987: 37). At 8:30 a.m., they heard on the radio that the rig had sunk: "[T]he companies had the information, they should have given it to the families first before it went on the radio. They had the information because they gave it to the radio stations." Brian Bursey remembers that the two eldest Bursey brothers drove to ODECO's office to try to get news about their youngest brother, Paul. "There wasn't much moving that morning—the roads were so full of snow. I figure it was around noontime that we made it to ODECO's offices, to try to get an answer from them as to what was going on out there. When we arrived, it looked to me as if they were all packed up and gone" (Bursey 2010). ODECO closed up shop, and Mobil Oil prepared a press conference. "They were all ready. In fact, they delayed the press conference so that my cameraman, John [O'Brien], could be sent home. His son worked on the rig and Mobil's people knew that this press conference was no place for a father. There would be no good news that morning," CBC reporter Marie Wadden remembers (2010).

On CBC's *The National* that night, a young *Ocean Ranger* worker, Robert St Aubin, described the confusion of a few days earlier, when the *Ocean Ranger* listed and the crew crowded around lifeboats on the rig they nicknamed the "*Ocean Danger*." Next came a clip of William Mason, president of Mobil Oil Canada, assuring viewers from a press conference at a St John's hotel: "We do watch things. To my knowledge the safety procedures were adequate" (CBC February 15, 1982). The piece continued with "angry questions in the House of Commons," the voiceover explaining that in the two weeks leading up to the disaster, "the provincial government has been scrambling trying to put in place some safety legislation." The CBC report closed with assurances that the province would hold a "lengthy investigation." The companies had neglected safety, our governments had let them, and everybody knew it.

Hope prolonged the effects of the disaster, as families followed the search waiting at least for a body to bury (Hawco 1983: 4). The public was shocked not only by the loss but also by the increasingly evident failure of the governments and oil companies to meet reasonable standards for safe operation. Newfoundland was traumatized and Canada was shocked. Over the next few days, twenty-two bodies from the *Ocean Ranger* were brought into St John's harbour. Coordinating the grisly task of identifying the bodies was Ray Hawco, formerly a Roman Catholic priest, then the Director of Community Relations for the Newfoundland and Labrador Petroleum Directorate, the provincial regulatory agency. The terrible storm also sank a Soviet ship, the *Mekhanik Tarasov*.[2] Thirteen bodies were recovered from the Soviet vessel, while five of its men were rescued from the waves.

Recovery crews were terrified that they would mix the bodies up, confusing "our men" with the Russians. Hawco took would-be identifiers to view the recovered corpses. The young manager from Schlumberger, my brother Jim's company, toured the bodies at around nine o'clock one night. All the bodies collected that day seemed to look Russian... all but one. That one was taller, fairer and slighter than the others. Hawco wonders if this manager thought it was Jim. The viewing ended with the decision to send all those bodies to Russia for burial. Hawco's phone rang at two that morning—it was the young Schlumberger manager again: "I am just not sure. I gotta go back. I gotta go back and take another look. I just couldn't face the families if I sent one of my men to Russia. I just couldn't face the families." Hawco calmed the young man. The phone rang again, two hours later; at four in the morning and the young manager repeated, "I gotta go back. I gotta take another look. What if I am wrong?" (Hawco 2008). That body went to Russia for burial.

Perry Morrison's father, Scotty, on February 16, 1982 (Courtesy of Canadian Press)

Perry Morrison was the youngest of eight children and a professional diver. He was trained to be in and under the water. If anyone had the skills to survive a little longer in the North Atlantic it was the divers. Scotty Morrison, Perry's father, the National Hockey League's vice-president of officiating, flew to St John's from Toronto as soon as he could. Like my brother Jim's, Perry Morrison's body was not recovered. Nor were sixty others. Morrison recalls the press conference when Mobil announced that the search for survivors had ended: "She just walked in with a stack of papers, dropped them on a table saying, 'Here are the names of the eighty-four men lost and presumed drowned.' Then she walked out. They were just cold. It was unbelievable and I will never forget it" (Morrison 2010).

CBC's *The National* reported on the 16th that there was no hope that anyone had survived and that no single agency was responsible for safety conditions on the rig. Three days after the loss, *The National* reported that the rig's lifeboats might have been useless, not ready to launch and inadequate to winter conditions in the North Atlantic. Scotty Morrison talked to Barbara Frum on *The Journal*, putting a nationally known face to the disaster. Morrison is an icon of

Canadian hockey. The youngest ever referee in the NHL, he went on to serve as referee-in-chief of the NHL from 1965 to 1988 and worked to preserve hockey's history in the Hockey Hall of Fame until his retirement in 1998. He describes his own induction into the Hockey Hall of Fame with a joke:

> "You were inducted into the Hockey Hall of Fame with that other referee, that Andy Von Hellemond." And I say, "That's right and there was another player that I can't quite remember..." Of course, that other player was Wayne Gretzky. And then I kid Andy, and I say, "Andy, you know what you and I are? We're the trivia question years from now, 'Who were the other two idiots who went in the Hall of Fame with Number 99?'" (2010)

GATHERING THE PIECES IN THE EARLY AFTERMATH

Chief Justice T. Alex Hickman recalls that he was hearing a violent rape case when an assistant came to the bench: "You have a phone call, your honour." Hickman was astonished. "I'm hearing a case!" The assistant said, "You'll want to take this call, I think: it's the prime minister." On the phone were Prime Minister Pierre Trudeau, Energy Minister Marc Lalonde and Attorney General and Justice Minister Jean Chrétien. "You have to do this inquiry" he remembers the prime minister saying (Hickman 2008). Not about to let a disaster get in the way of their dispute over the offshore, the Newfoundland government announced that it would conduct its own, parallel, public inquiry.

On February 19, 1982, flags were flown at half-mast at all federal buildings across Canada. In Newfoundland, Premier Peckford had declared a Day of Mourning, and two thousand Newfoundlanders gathered in the St John's Basilica to mourn the eighty-four men while thousands more watched the televised memorial service. The archbishop for the Diocese of Newfoundland, Alphonsus L. Penny, called from the pulpit for a joint federal and provincial inquiry. Archbishop Penny clearly recognized the incredibly divisive potential of this loss: fifty-six of the dead men were Newfoundlanders, and the rig was operating under inadequate regulation in waters that the provincial and federal governments disputed. If this was not another example of the mainland wanting all the benefits from Newfoundland's workers and natural resources without giving them any support, what was it? Penny also took the opportunity of his homily to advise the next of kin to be cautious in their dealings with the out-of-town lawyers who descended on the island. Meanwhile, Captain Carl Nehring hit the news with his story of having resigned as master of the *Ocean Ranger*, noting as possible contributing factors in the disaster numerous documented deficiencies in design, structure, chain of command and maintenance of the rig as well as Newfoundland's policy of forcing the oil companies to hire a quota of local workers (*Evening Telegram* 1982b). While mourning and anger mingled, the first lawsuit was filed by an American widow for two million dollars in damages in U.S. Federal Court in New Orleans (*Evening Telegram* 1982b).

T. Alex Hickman, Chief Justice of Newfoundland's Supreme Court Trial Division, was pronounced head of the federal inquiry. Hickman was a one-time contender for the provincial Liberal leadership. He lost to two other Newfoundland political legends: Joey Smallwood, who won, and John Crosbie, who came second. Hickman then sat as a Progressive Conservative member of the Newfoundland Legislature.[3] Having served as minister of justice and attorney general in Newfoundland as both a Liberal and a Tory, from 1966 to1969 and again from 1972 to 1979, along with stints as minister of health, intergovernmental affairs, finance and education, Hickman was a respected jurist, a slick politician with a public profile and an undoubtedly keen sense of how important it would be to take hold of the *Ocean Ranger* story before it turned against Newfoundland. As the MP for St John's East put it: "He knows better than anybody the political ramifications (of the inquiry)" (McGrath in the *Evening Telegram* 1982w). Hickman is described by a colleague as "one of the best legal minds in Canada" (Gold 2011).

An *Ocean Ranger* lifeboat, mangled almost beyond recognition, was televised on its arrival in St John's a week after the loss. Still, *The National* reported, "a date cannot be set for an inquiry into the *Ocean Ranger* disaster because the federal government and Newfoundland are squabbling over how the inquiry should be run" (1982b). It is a wonder that Prime Minister Trudeau and Premier Peckford could not at least agree that the *Ocean Ranger* disaster had the potential to create havoc and even to make everyone wonder why we were sinking so much money into oil exploration in such risky conditions.

On the eighth day after the disaster, discussion of a joint federal and provincial inquiry had descended into a "he said, he said" exchange of telex messages between federal Energy Minister Marc Lalonde and provincial Energy Minister Bill Marshall.[4] On the tenth day after the *Ocean Ranger* loss, the governments announced that there would be a joint federal and provincial inquiry into the disaster with Chief Justice Hickman as chair. The Newfoundland energy minister insisted that it was "not important" who suggested the joint inquiry, downplaying both the Catholic Church's enduring influence in the province and the bumbling and sniping between federal and provincial energy ministers (Doyle 1982).

The U.S. held Congressional hearings into the disaster, where ODECO's president, Hugh Kelly, testified. Kelly blamed the weather, the workers and providence. He blamed anything but ODECO. He laid out the main lines of

Ocean Ranger lifeboat
(courtesy of Mike Heffernan, originally printed in the Evening Telegram*)*

ODECO's communications strategy, emphasizing both the sublime combination of bad luck that led to the sinking and the human hands at the origin of the fatal chain of events.[5] "Stating that the giant rig had been praised after an informal inspection by a U.S. Coast Guard officer in training last October, Kelly observed that 'there are ways to sink any rig—there are ways to sink the Titanic.'" And "These people were all human beings and God knows what happened that night…. Whether we had a panic situation no one knows" and finally "It's hard to believe something 18 inches in diameter could start a chain of circumstances would end in such disaster" (*Daily News* 1982).

Meanwhile, the oil industry, as a community, reacted: Ted Byfield, publisher of the *Alberta Report,* wrote an impassioned letter insisting that the men must be remembered as heroes who sacrificed themselves to progress in the spirit of the sixteenth-century explorer Sir Humphrey Gilbert (*Alberta Report* 1982). Byfield warns the bereaved that the *Ocean Ranger*'s crew "may be prevented from dying as heroes. They may perish instead as mere victims, their lives claimed by rumoured negligence and suspected incompetence. This would be a worse tragedy than their deaths for it would eclipse the reason why they were there. That the Western world should find oil beneath the sea off Newfoundland is surely as important to its fate in the 20th Century as was the establishment of the first colony there to its fate in the 16th."[6] *Oilweek* took a more critical stance in its main news article on the disaster, voicing two questions: "What went wrong with a ship touted by her owners as unsinkable? And how, given the crew had at least 40 minutes warning, could the evacuation have been so badly bungled that not one man survived to relate the details of the catastrophe" (*Oilweek* 1982). *Oilweek*'s editor was cautious, anxious about a potential backlash: "There will be a tightening up as happened in Norway following the loss of the *Alexander Kielland* [a rig that sank in 1980 with 123 casualties]. However, it is to be hoped that there will not be a proliferation of new rules, regulations and new measures" (Humphries 1982).[7] The *Oil and Gas Journal* reassured readers that there was no oil spilling into Canadian waters from the well the *Ocean Ranger* was drilling. "'We know the well has a cement plug in it. We know the drillpipe was up and the hole was plugged,'" a Mobil spokesperson said (*Oil and Gas Journal* 1982a). Mobil Oil Canada published their newsletter with a photograph of a cross standing out against long summer grass and a big sky, and the first two pages featured a photograph of the *Ocean Ranger* faced by a poem—white print on black—that begins "Eternal Father, strong to save,/ Whose arm hath bound the restless wave." The newsletter's contents list explains that the cross is really the mast of a shipwreck on Sable Island. Other than that, the newsletter is run of the mill, though the guest editorial came from their legal counsel and emphasized that "the future for the oil and gas sector in Canada is full of promise" (Mobil Oil 1982). The oil industry positioned the disaster squarely in the long tradition of maritime fatalism: to go to sea is to place oneself in the hands of God.

Premier Peckford called an election one month to the day after the disaster. He is thought to have postponed calling an election in February at least in part because of the *Ocean Ranger* loss (*Newfoundland Herald* 1982a). The disaster played virtually no explicit role in the election campaign; in fact, "Premier Peckford and Energy Minister Marshall maintained a measured offensive in dealing with occasionally awkward questions from the media regarding safety inspections, the February 6 listing incident, etc." (O'Neill 1983: 160). Mere days after the disaster, Newfoundland politicians made speeches about building their own destiny while Canadian politicians counselled self-determination. For both, this meant a story of Newfoundland finally becoming a full partner in Confederation if it successfully managed its petroleum resources.

On the first day of the campaign, Peckford reasserted the promise of oil. He reminded Newfoundlanders of his determination to get a better deal on Upper Churchill. The Churchill Falls hydroelectric project was a symbol of the exploitation of Newfoundland and Labrador resources by mainlanders. Thousands of kilometres of Innu land were flooded, and Hydro Quebec negotiated its way to a sweet deal because the contract did not allow Newfoundland and

Labrador to raise the price. So when electricity prices surged, Hydro Quebec got a windfall and Newfoundland and Labrador got nothing (Cadigan 2009: 254–58). Peckford emphasized that he was "conscious of these bad deals of the past and how they have so severely hurt our people" and that his government "has proceeded with caution on the offshore." Past caution allowed optimism for the future. Peckford continued: "Jobs can be provided today if we begin now to work toward development of Hibernia. Over the next five or six years, preparation for this development can create thousands of jobs. We must make sure that as many of these jobs as possible go to people in this province" (Peckford 1982a: 4–5). Even as Peckford positioned himself as the key to the new Newfoundland, a column ran beside the tidings of new beginnings through oil, reporting the amounts of money donated from across Canada to the *Evening Telegram*'s Ocean Ranger Disaster Fund. John Crosbie called Peckford, "The Labrador Retriever," referring to the Premier's determination to retrieve hydro resources that were sold cheap to Quebec. "He's the Newfoundland dog," while "the Liberal party is a scared poodle" (*Evening Telegram* 1982g). Peckford's central leaflet concluded its impassioned plea to voters: "Now, with jobs and future security depending on settlement of the offshore, I urge you to join me in telling Ottawa that we are not prepared to wait for our future. We need jobs now and we need to plan. On April 6th stand with me for the future we've already begun" (NLPC 1982).

The *Ocean Ranger* disaster was simply written out of the earliest version of this "already begun" future. Campaigning amid the grief and amorphous anger of the disaster's immediate aftermath, Peckford played on longstanding suspicion of the federal government and easily deflected attention away from the provincial government's negligence and failure to regulate health and safety in the offshore. "We are involved in a revolution," Peckford declared (in O'Neill 1988: 271). He elaborated: "It's a revolution between the ears, a great leap forward in attitude. There's a change taking place, young people and old people saying, 'Holy smoke, we did it wrong in the past. Now we've got a chance to make a new start'" (*Evening Telegram* 1982h). "If standing up for Newfoundland's right to be equal is considered to be making war, then boys, I'm waging war," Peckford blustered, reclaiming the Liberals' slogan of "Make Work, Not War," to reinforce his own image as the "fighting Newfoundlander" (*Montreal Gazette* 1982). To disagree with Peckford's approach to negotiating for a share in control of the revenues of the offshore was to be a traitor. As Peckford said on the last weekend of the campaign: "They're traitors, traitors to Newfoundland ever having a fair chance at the future" (*Leader Post* 1982a). Peckford ran against Ottawa and he defined the voters' choice: they could return to a past of colonial exploitation or they could embrace a future of liberation through offshore development. His main platform plank was his "blueprint for development of Newfoundland's offshore oil reserves" (*Newfoundland Herald* 1982a). The *Ocean Ranger* disaster sharpened Newfoundlanders' sense of themselves as a colony exploited by outsiders, which gave credence to Peckford's claim to "stand up" for the province against Ottawa's predation. "Premier Brian Peckford hopes to arouse Newfoundlanders to vote emotionally on a 'them or us' issue. He terms it a 'kind of referendum, in Newfoundland terms' with the choices being a vote for Newfoundland or a vote for '100 percent (Federal) control of our oil and gas resources'" (*Newfoundland Herald* 1982b). It is a testament to the groundwork laid in communicating the promise of oil that the villainous strangers were believed to be from Ottawa and not from New York, New Orleans and Houston, the hometowns of the oil companies. That Peckford's government failed to enforce existing regulations, let alone to develop new ones for the promising industry, did not resonate with Newfoundland voters. The Peckford government won a landslide re-election: of the total of fifty-two seats, in 1979 the Liberals won nineteen seats while the PCs won thirty-three; in 1982 the Liberals won only eight seats while the PCs swept forty-four.

Three months after the loss of the *Ocean Ranger,* on May 17, 1982, Peckford declared another official day of mourning, complete with black arm bands. This time the premier encouraged Newfoundlanders to mourn a blow to their sovereignty: Prime Minister Trudeau's referral of the offshore dispute to the Canadian Supreme Court. "It is an action that one would only expect from a foreign, hostile power and not one from a national government which is supposed to protect and nurture equality and justice throughout the land," Peckford said (*Leader Post* 1982b). The *Ocean Ranger* disaster and the assertion of the federal right to have the question of jurisdiction heard in a Canadian court were somehow on a par: both threatened Newfoundland's emerging sense of sovereignty. Both events affirmed collective identity in the face of colonial exploitation by mainlanders—a hallmark of successful Newfoundland politics.

In 1982, Newfoundland's offshore was still only promising, but both provincial and federal governments had invested heavily to convince their constituents that oil was the key to a new beginning. For workers and their families, the promise of oil meant an exchange of work for pay, conducted under what they assumed were reasonably safe conditions. Everybody knew that the oil would be worth exponentially more than it would cost the companies to find it and siphon it from under the sea. Everyone also knew that, as T. Alex Hickman put it, "Nobody ever goes to sea not realizing that the sea is master of your fate" (2008). The sense of betrayal that emerged in the disaster's aftermath did not come from a sudden discovery that companies seek profits or that the ocean can kill. The shock was in governments' and oil companies' failure to acknowledge that a rig is a ship on which people live and work, not just a platform drilling for oil.

Before the benefits started to flow from under the ocean floor into St John's, investing in offshore development seemed credible only when people were convinced that governments could make the oil companies work, at least in part, for the interests of Newfoundlanders and Canadians. The *Ocean Ranger* disaster showed that, at least in 1982, governments failed to protect people's very lives, let alone the public's economic interests. The federal Liberal government's attempt to keep petroleum profits in Canada through the National Energy Program had seriously irked international oil interests (Laxer 2008). The Peckford government was positioning itself as the path to self-governance: Peckford could work with the oil companies to bring Newfoundland to maturity. The disaster threatened to remind people that the benefits of oil development depended on governments' abilities to harness an industry that was fundamentally indifferent to local interests.

For a few hours in 1982, government negligence and corporate predation became the top news story, and the shock threatened to reawaken the fatalism and insecurity of the province's colonial past. As the former executive administrator of the Ocean Ranger Families Foundation put it: "Newfoundland was shaken to its roots by the sinking of the *Ocean Ranger*. The government had us convinced that the oil industry was the miracle cure and the start of a new era. We had this piece of technology out there, a veritable fortress, the unsinkable rig, but then it sank and all these people died. In a cultural sense, so did our dream" (Newhook, in Heffernan 2009: 181).

PERSONAL AND COLLECTIVE TRAUMA

Politicians presented the promise of oil to the people of Canada and Newfoundland as an investment in the future, a way out of dependency and towards participation in the most lucrative industry in the world. The *Ocean Ranger* story had to be managed if Newfoundland was to realize its destiny through the offshore. Governments and oil companies needed to appear to have the same interests as communities, workers and families. They needed to show themselves "learning a lesson," relegating the *Ocean Ranger* disaster to a time "before" Newfoundland grew up—a time of passivity and subjugation to the ocean, to colonial feelings of inferiority and to exploitative outsiders. In the 1970s and early 1980s, Newfoundland's cultural renewal was underway, as evident in the emergence of the Wonderful Grand Band, Codco, the growing interest in Newfoundland folklore and the development of the *Dictionary of Newfoundland English*. Newfoundland's full economic surge was still but a politician's promise: if we trust in oil, Newfoundland can grow up, "breaking the cycle of dependency," as Prime Minister Brian Mulroney put it when the federal and provincial governments finally signed the Atlantic Accord on February 11, 1985. Four days before the third anniversary of the *Ocean Ranger* disaster, the press packages heralding the historic accord included a timeline of oil development on Newfoundland's offshore, with one glaring omission: the loss of the semi-submersible oil rig the *Ocean Ranger* and its crew of eighty-four men (Newhook 2008).

By 2007, when filmmaker Mary Sexton had a script, crew and a cast lined up to shoot *Atlantic Blue,* a feature film based on the *Ocean Ranger,* its title drawn from Ron Hynes' popular song commemorating the loss, no oil company would let her shoot on any of the available rigs. "This was not an exposé on big oil, this was about a community. It was like the Klondike for us," Sexton says. "We really felt like this was going to turn us around, this was going to be the jobs and when that disaster happened it wasn't just the men, it was the mothers, it was the wives... Everybody knew somebody that was on that rig and it just basically polarised us" (*Telegram* 2007). The memory of the disaster is often managed more subtly, as in the excellent 2002 documentary from Gzowski Films which demonstrates Mobil and ODECO's negligence and mismanagement only to conclude with authoritative fatalism in a voiceover from Newfoundland icon Gordon Pinsent: "The *Ocean Ranger* disaster has taught us that neither our most advanced technology, nor our grandest dreams, can conquer the power of the sea."

Today, the pulse of oil beats strongly in St John's, in the boutiques and bars, galleries and high-end restaurants. St John's has the "hottest job market in the country," with employers complaining that they cannot find workers, especially in the petroleum industry and its spin offs (Moore and Grant 2011). The wrangles between Premier Peckford and Prime Minister Trudeau for jurisdiction over the offshore have faded from memory, or more accurately, they have been transformed into an historical past, along with the *Ocean Ranger* disaster.

Despite outright denial and seafaring fatalism, the *Ocean Ranger* loss haunts today's flourishing oil industry in Newfoundland. On the eve of becoming Newfoundland and Labrador's lieutenant governor in 2008, John Crosbie lashed out at both federal and provincial governments for failing to follow through on what he saw as one of the key recommendations of the Royal Commission. Citing the Cullen Report on the *Piper Alpha* sinking of 1988 as a clear vindication of Hickman's findings, Crosbie wrote: "Canadians must insist the government and the offshore industry immediately carry out the recommendations of the 1984 commission. The indifference and slothfulness of Canada and its bureaucracy, especially in Transport, and at the CNLOPB [Canada and Newfoundland-Labrador Offshore Petroleum Board] should not be tolerated another day" (2007). Among the most powerful and prominent Newfoundlanders of his generation, John Crosbie was marked by the *Ocean Ranger* disaster and by his failure to see all of the recommendations of the Royal Commission fulfilled. "We still don't know how to get men off those rigs," Crosbie repeated angrily in a 2008 interview.

Retired Chief Justice T. Alex Hickman choked up when he told me in 2008 about the heroism of the supply boat workers as they struggled, in unimaginable seas and weather, with no training and inappropriate equipment, to try to save some of the *Ocean Ranger*'s crew. Howard Pike of the CNLOPB gave me one of his two reassuringly well-worn copies of the report of the Royal Commission inquiry into the disaster. A commemorative plaque hangs in the waiting room of the CNLOPB, and everyone who passes through that office sees an etched image of the rig and the names of the eighty-four dead. In the hallway into the inner sanctum of CNLOPB there is a painting of the memorial to the men. The memorial sculpture itself stands on the grounds of the Confederation Building, home of the Newfoundland and Labrador House of Assembly. Taped to the wall of the Marine Institute where all rig workers now take their safety training is an old copy of the St John's daily newspaper story about the twenty-fifth anniversary of the *Ocean Ranger* disaster with a photograph of the rig, the capsized lifeboat and Gary Wall, the "85th Man," that is, the last man from the rig to land on shore alive.

As for me, every year, around the time the heart-shaped boxes of chocolates appear in the stores at the beginning of February, I get anxious about the *Ocean Ranger* anniversary. My anxiety builds as the fifteenth approaches. The anniversary of the disaster is often the day we teach Karl Marx in the University of King's College Foundation Year Programme, where I am now a professor of the humanities. Marx says: "Men make their own history but not of their own free will; not under circumstances they themselves have chosen but under given and inherited circumstances with which they are directly confronted" (1996: 36). I look at the *Ocean Ranger* disaster from my perspective as a family member and as a political theorist whose education was subsidized with "blood money" from the settlement the companies paid to my family.

My brother's death in an oil rig disaster made me a member of an international community of mourners and people whose confidence in legal, economic and political authority was forever marked by the failure of Canada and Newfoundland to regulate those oil companies. My family is an "*Ocean Ranger* family" though we are not Newfoundlanders. This gives me what anthropologists call "insider-outsider status"; I am part of the community but not fully identified with it. Readers looking for new facts about the causes of the disaster or an assessment of the current safety regime on Canada's offshore will have to

THE PEOPLE'S PAPER

The Telegram

ECMAs kick off in Halifax Page B4

Fog Devils outlast Cataractes Page C1

www.thetelegram.com

St. John's, Newfoundland and Labrador

Thursday, February 15, 2007, Vol. 128 • No. 311

25 years ago, a dark day dawned

On Feb. 15, 1982, the Ocean Ranger (shown above) sank off Newfoundland's coast in a fierce winter storm. Eighty-four men lost their lives in the disaster, the worst in the history of Newfoundland's offshore oil industry and one of the worst in provincial history. — Photo by The Canadian Press

So many lives lost

The Ocean Ranger capsized and sank during a winter storm that lashed the Grand Banks on Feb. 15, 1982.

Eighty-four men died that night.

Following are the names of the 84 men lost, as listed in the Royal Commission on the Ocean Ranger Marine Disaster report. Fifty-six were Newfoundlanders.

Robert Arsenault, St. John's
George Augot, Torbay
Nicholas Baldwin, Carbonear
Kenneth Blackmore, Norris Arm
Thomas Blevins, Plainfield, Conn.
David Boutcher, Corner Brook
Wade Brinston, Arnold's Cove
Joseph Burry, Norman's Cove
Paul Bursey, St. John's
Greg Caines, St. John's
Kenneth Chafe, Topsail
David Chalmers, St. John's
Gerald Clarke, St. John's
Daniel Conway, St. John's
Gary Crawford, St. John's
Arthur Dagg, St. John's
Norman Dawe, Harbour Grace
Jim Dodd, Berwick, N.S.
Thomas Donlon, Sumter, S.C.
Wayne Drake, Kilbride
David Leon Droddy, Covington, La.
William Dugas, Abbeville, La.
Terrance Dwyer, Carbonear
Domenic Dyke, Eastport
Derek Escott, Mount Pearl
Andrew Evoy, Mount Carmel
Robert Fenez, St. John's
Randell Ferguson, Natchez, Miss.
Peter Fogg, Mount Pearl
Ronald Foley, St. John's
Melvin Freid, St. John's
Carl Fry, St. John's
George Gandy, Logansport, La.
Guy Gerbeau, Montreal, Que.
Cyril Greene, Piccadilly
Reginald Gorum, El Paso, Tex.
Norman Halliday, Toronto
Fred Hamum, St. John's
Tom Hatfield, Wolfville, N.S.
Capt. Clarence Hauss, Baltimore, Md.
Ron Heffernan, St. John's
Gregory Hickey, Torbay
Robert Hicks, Goose Creek, S.C.
Derek Holden, Mount Pearl
Albert Howell, Mount Pearl
Robert Howell, London, Ont.
Robert C. Howland, Calgary, Alta.
Jack Jacobsen, Tusket, N.S.
Cliff Kuhl, Brooks, Alta.
Harold LeDrew, Botwood
Robert LeDrew, Botwood
Robert Madden, Calgary, Alta.
Michael Maurice, St. John's
Ralph Melendy, St. John's
Wayne Miller, St. John's
Gord Mitchell, Lacombe, Alta.
Perry Morrison, Weston, Ont.
Randy Noseworthy, St. John's
Kenneth O'Brien, St. John's
Paschal J. O'Neill, Ferryland
George Palmer, St. John's
Clyde Parsons, Foxtrap
Donald Piercey, Bishop's Falls
John Pinhorn, St. John's
Willie Powell, Franklinton, La.
Gerald Power, St. John's
Douglas Putt, Goulds
Donald Rathbun, Narragansett, R.I.
Darryl Reid, Upper Gullies
Dennis Ryan, Medford, Ore.
Rick Sheppard, St. John's
Frank Smit, Kilbride
William Smith, St. John's
William David Smith, Valley Station, Ky.
Ted Stapleton, Mount Pearl
Greg Tiller, Mount Pearl
Craig Tilley, St. John's
Benjamin Kent Thompson, Hattiesburg, Miss.
Gerald Vaughn, Collins, Miss.
Woodrow Warford, Carbonear
Michael Watkins, New Orleans, La.
Robert Winsor, Calgary, Alta.
Robert Winsor, Paradise
Stephen Winsor, Paradise

The 85th man

A disaster at home saved Gary Wall from a much bigger disaster at sea — the sinking of the Ocean Ranger

By Moira Baird
The Telegram

Gary Wall avoided one of Newfoundland's worst marine disasters 25 years ago when he left the Ocean Ranger to deal with his own disaster, a house fire.

Wall was flown off the doomed rig just 24 hours before it capsized and sank in the early morning hours of Feb. 15, 1982.

The remaining crew of 84 perished.

"A day don't go by when I don't think about the Ocean Ranger — not a day," said Wall.

"I can close my eyes and still see it now. It was a really nice night. It was windy — not a cloud in the sky.

"I always liked to get in the (helicopter's) back seat, and I remember looking back out the side window at the three rigs ... I can still see that in my mind when I close my eyes."

Wall was the rig's ...

He doesn't talk about it much.

The RCMP took a statement from him a couple of days after the rig sank.

They asked questions about how the rig was stabilized, the ... what job, and whether he had witnessed any problems.

"It was a pretty quick interview," said Wall. "I was upset at the time."

An overturned lifeboat was all that was left 25 years ago Feb. 14, when the Ocean Ranger drill rig capsized in a violent storm, killing 84 men. — Photo by The Canadian Press

He wasn't called to testify at the royal commission hearings, and has given one media interview — five years ago.

Wall says the Ocean Ranger was often described by the bosses as the newest, best rig in the world.

"That was part of their problem — this is my guess, now.

"The Americans are strong-headed, or bull-headed people — they would drill till the last minute.

"They would push right to the edge, and that's what happened out there. They drilled too long, they got caught in the storm and they thought they were invincible — this rig is not going to sink — and it caught up on them."

Wall started working in the offshore in 1979 for a variety of rigs drilling exploration wells on the Grand Banks.

His favourite rig was the John Shaw, a small semi-submersible that Wall says "bounced like a cork."

He always felt safe aboard that rig.

See 85TH, page A4

INDEX

WEATHER

88¢

PROVINCIAL

WWW.THETELEGRAM.COM

Haunted by the memories

Former Ocean Ranger worker recalls the many friends he lost

Ocean Ranger commemorations

One man's testimony

As the radio operator, you were there with the guys when they called their wives or when they were sick.

Lloyd Major

Not scheduled to work

Fateful chronology

Newspaper story about the 25th anniversary on the wall of the Marine Institute (photo by Gord Graham) look elsewhere. My focus is on the ways the disastrous event is remembered, and how these obscure what I see as the original problem, namely, that governments thought their constituents were best served by giving corporations a free hand.

I had just turned sixteen when Jim and his eighty-three co-workers died on the *Ocean Ranger*. I cannot remember the first days very well. I have impressions of the hushed house, of everyone's nerves wired to the telephone and of the terrible decision to "call the boys" and summon my other three brothers home for a funeral. Until I interviewed Pat Ryan, whose son Craig Tilley also died that night, I had the impression that the search for survivors had gone on for about four days. "No, dear, we knew that day," she corrected me.

Trauma fragments our sense of self and it alters our relation to time. Scotty Morrison, whose son Perry was lost in the disaster, recalled: "The craziest thing is … I got a call from … what was the company Perry worked for? Hydrospace. Hydrospace. 4:28 in the morning, 'Mr Morrison I just want to advise you before you hear it on the news that the *Ocean Ranger* has capsized.' I'll bet you for two months I woke up at 4:28 in the morning." When individuals suffer a shock, a blow to the body or a sudden terror, the mind becomes vulnerable, violated and anxiously scans the horizon for the next threat. Sociologist of disaster Kai Erikson describes trauma: "Something alien breaks in on you, smashing through whatever barriers your mind has set up as a line of defence" (1995: 183).

Trauma makes us strangers to ourselves and to the world, and we repeat the known hurt in both literal and symbolically coded replays of the traumatic event. In the first months after the disaster, I had two recurring nightmares: one a very literal-seeming dream of the men trying to get off the rig in the storm. This dream cannot be factually accurate because I know nothing about rigs or evacuations or even hurricane-whipped seas. And yet that dream is in some ways my "truest" memory of that night. My other nightmare was more symbolically coded: I was in my bed in the house in Berwick and I woke to find the house dismantling itself around me, silently. When the walls fell away, I floated in a void in my bed. Then the bed peeled off and floated away. Finally, I was left struggling to keep my blankets around me.

When the nightmares stopped, I stopped thinking about that "evacuation." I realized this almost thirty years after the disaster when I visited the memorial sculpture at the Confederation Building in St John's. As I touched the raised names of the men on the plaque that is nestled in the enclosed garden, it struck me that I had literally not let my self think of those last two hours for decades. I met Lisa Moore and Mike Heffernan around this time, too, and read both *February*, Moore's novel about the widow of a man lost on the *Ocean Ranger*, and *Rig: An Oral History of the Ocean Ranger Disaster* in manuscript form. Both of those works return to the most intimate moments of the disaster. For me, personally, and it seemed for Newfoundland as a community, a period of latency had ended; aspects of our memories that were suppressed for decades were now for some reason accessible. We were ready, in our own ways and from our own vantage points, to retell the story of the *Ocean Ranger* in our own words.

Relations between individual and collective traumas are complex, and they are played out in a web of interrelated ways of remembering and describing what happened. Writing about my own trauma as it interacts with a collective trauma requires an interdisciplinary approach, and I use a mix of writing styles and reading strategies. I consider public practices of remembering and relate them to my own memory and the remembrances of some other family members and key public figures of the *Ocean Ranger* story. I interviewed a few other family members in order to enrich my own understanding and to gain some insight into a wider range of experiences. My overarching question is this: How do our recoveries from personal and collective trauma relate to the capacity of liberal capitalism to stave off crises of confidence in our political and economic systems?

In the first chapter, I look at the Royal Commission of inquiry and I consider not only what it "found" in terms of the causes of the deaths and the loss of the rig, but also, in a broader socio-political sense what the inquiry process and its reports contributed to "the" *Ocean Ranger* story as a moment of historical memory. In the next chapter, I tell the story of some family members' financial settlements with the oil companies. Then I consider the history of such "blood money," tracing its roots and stigma deep into our cultural past. The inquiry and the financial settlements are ambivalent: they realize some aspects of justice while they suppress others. Together, they compensate for the failure of Canadian criminal law to police corporate crime in that they divert the vengeance-energy of families and they deflect political criticism. In the fourth chapter, I take a broad look at practices of commemoration. From the earliest service at St John's Basilica through to Mike Heffernan's and Lisa Moore's retellings of the disaster, Newfoundland and Canada have commemorated this tragedy in diverse ways, all of which contribute to our collective memory. In the fifth chapter I draw these together in a map of the "aftermath of industrial disaster," and in the sixth chapter I develop this map into a method for interpreting the politics of memory, particularly the tensions between liberal capitalism's capacity to manage legitimation crises and the unpredictable character of collective trauma. The fifth and sixth chapters will be of most interest to readers who are interested in ideology critique and collective trauma. Readers who are not interested in these more theoretical matters are welcomed at any point to turn directly to my conclusion. In the book's final chapter I look at how the aftermath we experienced with the *Ocean Ranger* is being replayed in some ways by families and communities in the Gulf Coast in the wake of the *Deepwater Horizon* disaster that killed eleven men and triggered massive ecological devastation in 2010. The promise of oil is re-affirmed by the people of the Gulf Coast and of America more broadly despite the fact that the profit imperative of BP and Transocean led those managers to neglect safety and to ignore that a spill of that magnitude was probable. That no rig operating in the Gulf of Mexico had a real spill containment and clean-up plan should not surprise us because the governments left the companies to self-regulate (Safina 2011).

The processes that manage potential legitimation crises are always already at work in the day-to-day affairs of liberal capitalism, but a disaster that rattles the very bonds of community can make people more aware of them. The freer and more privileged the community, the harder such processes have to work to re-affirm public confidence in, or at least complacency about, the cosy relationship between regulators and the industries they are supposed to monitor and correct.

People often talk about needing "closure" in the wake of a traumatic loss, especially when that loss involves injustice or a betrayal of trust. "Closure" is of course a term that comes out of psychoanalysis, as the healthy outcome of a process of "working through." To achieve closure by working through the trauma, we make a traumatic event our own by revisiting and retelling or remodeling it. Closure brings a safe kind of remembering, one that allows us to revisit the traumatic past without being overwhelmed and controlled by it. It is a remembering that is possible only because we forget or at least contain the most violent kinds of memories. I return to this in my final chapters on commemorations and the map of a disaster's aftermath. For now, let me distinguish between the closure that comes from serious reflection on the event and its political and symbolic implications from a false closure that suppresses the most disturbing aspects of the event. As social and political theorist Theodor Adorno cautioned in the wake of World War II, we must be wary of glib accounts wherein, as he puts it, "'working through the past' does not mean seriously working upon the past"; instead, we must strive for "a lucid consciousness" the breaks the past's power to keep us hostage (1998 [1959]: 89). To demand closure too quickly, either as individuals or as a community, is, Adorno warned, "to close the books on the past and, if possible, even remove it from memory."

The "promise of oil" is a promise of infinite development, constantly growing wealth, of collective identities defined by their ability to wrest recognition and benefits from transnational companies whose interests coincide only accidentally with their host communities. "The people of Newfoundland" grew to maturity by standing up to the oil industry, so the story goes. While the local benefits of oil development are undeniable, the preventable crash of

Cougar Flight 491 and the *Deepwater Horizon* disaster should remind us that without regulation international companies will not "naturally" work in our best interests. And governments will not regulate unless "the public" demands that they do so.

NOTES

1. Between 1978 and 1982, oil companies paid only an average of 19 percent of overall costs of exploration in Atlantic Canada (Canadian Tax Foundation, in O'Neill 1988).
2. The operators of the *Mekhanik Tarasov* were sued successfully for failing to take appropriate measures to safeguard against the clearly forecast storm. This suit was appealed and the decision overturned.
3. It is best for outsiders to keep in mind that however inadequate the political spectrum running from left to right may be in other parts of Canada, in Newfoundland (as in Quebec), the situation is even less clear. Both forms of provincial nationalism have strong socialistic, paternalistic and capitalistic elements that combine in unique ways. Newfoundland does, of course, have a "distinctive culture," and Williams followed Peckford ideologically, particularly in his determination to develop private enterprise while building "a strong sense of identity, pride and self confidence" (Speech from the Throne 2007, in Marland 2010: 10).
4. "Late Monday, the *Telegram* received a copy of a Telex that Lalonde had just sent to Marshall concerning this matter. The federal minister told Marshall he wished to correct certain impressions left and 'misstatements' contained in his recent Telex. Lalonde said he had asked Marshall on Feb 15 for his general comments regarding the federal inquiry but 'I received no such comments. Instead you notified me the next day that the Government of Newfoundland intended to proceed with its own inquiry and advised as to terms of reference'. Lalonde went on to say that, 'I am surprised and shocked that you suggest that I never offered a joint inquiry." For his part Marshall sustained the offensive saying that the matter is "a little too serious for these persons to be playing games with" (*Evening Telegram* 1982d).
5. This is both a public relations move and a nod to the tradition under common law, where "an injured worker was precluded from succeeding in a claim for damages where he was injured through the negligence of another workman. The employer could rely on the defences of "common deployment" (which relieved the employer from vicarious liability for accidents caused by the negligence of a worker's fellow servants), "voluntary assumption of risk" and "contributory negligence" (which provided a complete bar to the action until the introduction of apportionment or contributory negligence legislation) (Hayashi 1983: 167).
6. Bayfield continues: "If the crew of the Squirrel (Sir Humphrey Gilbert's ship that sank in 1583 after taking possession of Newfoundland for the Queen) may be said to have died gloriously, so too therefore did the crew of the *Ocean Ranger*, and their wives, children and parents have a right to know it." The author goes on to describe the fury of the storm as he imagines it. Gilbert is an important cultural

symbol, reportedly having died with the words, "We are as near to Heaven by sea as by land" on his lips (Earle 1998: 91).

7. He continues: "beyond those that may be needed in the interests of safety and safe operation. The disaster, while unlikely to affect the pace of exploration drilling, will obviously prompt far more attention being paid to the design and construction of production facilities in these waters."

1

“THE” *OCEAN RANGER* STORY: THE ROYAL COMMISSION

“Had the crew only… ”

We lived in Nova Scotia, and so we were not battered by the storm of media coverage of the inquiry testimony and the threatened lawsuits. My first memory of the Royal Commission of inquiry into the disaster is the appearance of the report at our house in Berwick the summer after my first year of university. I was eighteen, the age Jim had been when he left university and moved to the unheated cottage with no running water to “find himself,” as my father says; in other words, to drink, smoke, read, hang out with friends and figure out which direction to take in a job market that seemed to be closed them, as youth. I hid from my parents as I puzzled through the technical details about the ballast control, the power struggles among rig managers, the reconstruction of the workers’ misinformed responses to the deteriorating situation and the silence that followed their final radio message. Perry Morrison’s father, Scotty may have spared his family one of the aftershocks of the disaster with his refusal to have the report in his house. “Susan, I threw it away. I thought, ‘What a bunch of crap.’ Excuse the language. And I thought, ‘I don’t want this thing around.’ Joan, she didn’t want to read it and I didn’t want any of the kids to read it. So I just tore it up and took it to the dump,” he told me. Morrison refused a false closure the report offered. For me, as for Morrison, the inquiry report accomplished something very important, and yet its answers to the key questions fell short of the demands of justice. In fact, the report offered apparent closure in part by deflecting cultural dialogue away from the pursuit of culprits and towards the reconciliation of the local community with the international oil industry. The inquiry and its report did this by revising the story of regulatory failure and corporate neglect into a tale of innovative risk-taking that cost dearly but was redeemed by its contribution to technological, economic and cultural progress.

THE WORK OF A PUBLIC INQUIRY

The night after the loss of the men and the rig, CBC television reported that in the two weeks before the disaster, the provincial government had been "scrambling" to get safety legislation in place. In the House of Commons, Members of the Opposition demanded to know if there would be any criminal charges. Minister of Energy, Mines and Resources Marc Lalonde responded: "If there is critical liability or criminal liability which might be involved, I assume that this will be the subject of the conclusion by that commission of inquiry" (1982). He said this despite common knowledge that standards for evidence and powers to punish increase together: a public inquiry considers a wide range of evidence precisely because it has no power to punish, whereas a criminal trial accepts a relatively narrow range of evidence that is tightly procedurally bound so as to avoid wrongful convictions. Justice Hickman was clear also that the broad mandate of the inquiry was possible precisely because there would be no criminal charges (2008). Ten days after the loss, Newfoundland Solicitor General Robert Kaplan acknowledged that the RCMP were investigating but only as a matter of procedural necessity. Their investigation was not "'triggered by suspicions or by reports' of criminal negligence" he said (*Evening Telegram* 1982e). No criminal charges were laid, due in part to jurisdictional confusion and in part to Canadian law's weakness in holding corporations and their agents accountable.[1]

Chief Justice T. Alex Hickman accepted the job of presiding over the inquiry on condition that he have a say in its terms of reference and input into the selection of commissioners. He felt that Newfoundland's government, under Premier Brian Peckford, distrusted Canadian Prime Minister Pierre Trudeau and did not understand the limits of its constitutional rights. A provincial inquiry would have been restricted to reviewing the training program; jurisdiction over oil production was squarely a federal right and responsibility. Hickman's control was threatened when St. John's Archbishop Alphonsus L. Penny called from the pulpit for a joint federal-provincial inquiry. Hickman and the federal Liberals withstood provincial pressure to appoint a co-chair; in the end Hickman was the sole chair of the panel that included three provincially appointed and three federally appointed commissioners (Hickman 2008).

The formal mandate of the Royal Commission had two parts: to inquire into and report on the loss of all members of the crew of the *Ocean Ranger* and to inquire into a wide range of inspection, licensing, certification, search and rescue, pollution prevention and division of authority issues including "both the marine and drilling aspects of practices and procedures in respect of offshore drilling operations on the Continental Shelf off Newfoundland and Labrador" (RC 165–66). The inquiry was thus a combination of an accident inquiry, which focused on forensic explanation of a particular mishap, and a policy inquiry, which advised the governments on a wide range of policy possibilities for the emerging industry. In a broader, socio-political sense the inquiry addressed three crucial questions: What happened? Who is to blame? What is to be done? The goal was to lay out all the evidence in a quasi-judicial performance and then turn to the future as smoothly as possible. Culprits faced the community they injured, and they were cross-examined and shamed. But they did this without facing punishment.

Justice Hickman clearly understood the political and social importance of the hearings and findings of the inquiry, and he distinguished his commission's work from the more narrowly forensic work of the inquiries conducted by the American Coast Guard and the U.S. National Transport Safety Board. The U.S. Coast Guard inquiry into the *Ocean Ranger* sinking was important because it had already "discovered" the broad strokes of "what happened," which might have stolen some thunder from Hickman's inquiry. But, Hickman insisted, the goals of the inquiries were radically different. The scope of the Canada-Newfoundland inquiry was much broader than the American inquiries, and it entailed social as well as forensic and policy importance: "All the Coast Guard wanted to establish was that the Coast Guard had done an inspection and then did not follow up. Theirs was quasi-military: they were inquiring into their own conduct" (Hickman 2008). For Hickman, the *Ocean Ranger* inquiry was less controversial "in the true sense" than his own later inquiry into the wrongful conviction of Donald Marshall Junior in Nova Scotia would be. The Marshall inquiry was controversial "in the true sense" because it exposed systemic bias against Aboriginal people in Nova Scotia's judicial system (Hickman 2008). The impact of the Donald Marshall inquiry on the judicial system of Nova Scotia and Canada ranged from the personal (there are still members of the Nova Scotia legal establishment who cross the street when they see Justice Hickman) to the procedural: the Marshall inquiry altered standards of disclosure in criminal trials so that now there must be complete disclosure from the Crown to defence lawyers.

The joint Canada-Newfoundland Royal Commission was a first responder on the scene of the potential loss of confidence in the promise that offshore development was Newfoundland's ticket out of its colonial past. Itself born out of conflict over jurisdiction, the inquiry took shape out of fear of the increasing exposure of an even more fundamental conflict lying at the disaster's heart: the conflict between the rig workers' lives, Newfoundland's maritime culture and the profit drive of international companies.

Keeping up with the testimony, either in person or through daily media coverage, involved the inquiry's audience directly in recalling the events of that terrible night and of piecing together the mass of evidence into a story that explained how the shoddy operation of the rig had been allowed in the first place. CBC reporter Marie Wadden covered the disaster and the inquiries. She saw the *Ocean Ranger* disaster and its aftermath always through the eyes of her cameraman, John O'Brien, who lost his son on the rig. She also saw through the eyes of a woman journalist making her name by developing an expertise in an emerging industry. "In those days, nobody covered the offshore. So I thought, 'there's a space for me.' And then the *Ranger* sank and I covered that story for a couple years solid" (Wadden 2010). Wadden's files from the hearings of the Royal Commission convey the grind of the testimony. Day after brutal day, witnesses laid out a horror story of neglect, miscommunication, personality conflicts and technical incompetence. Some family members attended the hearings, though many did not. Greg Hickey's mother, Patricia, recalled:

> I did attend quite a few [days of the hearings] and a lot of it was technical which I didn't understand but a lot of it was very emotional. That was when all the men from supply boats testified and gave their version and especially the one with the crew from the *Seaforth Highlander*. We just sat there and listened and we were just absolutely awe struck. The way those men worked that night, it was just awful, it was like your mind could not grasp the condition that they had to work under to try to save some of those lives when their own lives were in danger as well. (Hickey 1984)

Following the testimony took its toll on all observers, of course, particularly because the lines between family and friends of the dead men, professional observers and the general public were blurred in the tightly knit community. As a diversion, maybe from the intensity, Wadden shifted to writing with her right hand (she's left-handed) for her note taking during testimony from one of the most notorious figures in the *Ocean Ranger* story, ODECO's drilling supervisor, Jimmy Counts. The report gives a relatively bland account of Counts' character as experienced by his employees, though workers in Heffernan's *Rig* would flesh that character out much more fully years later, as we will see in Chapter Four.

Royal commissions of inquiry are much more than simple fact-finding processes (if such things exist), and they are not simply ideological tools that governments use to sweep uncomfortable truths out of public debate. They are a little of both—fact-finder and whitewash—and much more besides. Royal commissions are a mix of legal process, social science and technical study. Generally speaking, such inquiries either address a policy question that the government wants comprehensively reviewed, or they study a particular event, often a disaster or a particular incident of extreme injustice, like Donald

Marshall's wrongful conviction and the repeated failures of the appeal system in his case.

Royal commissions began in their current form as a response to the problem of poverty in England. Marx famously pored over inquiry reports in the British Library, and his collaborator Frederick Engels wrote his own alternative to the official reports with his, "Report on the Condition of the English Working Class 1844." Inquires and their reports arose with the need to manage populations in the industrial revolution. Inquiry reports are notoriously frustrating for those who hope for roadmaps for policy improvement only to see the public bored into forgetful submission with technological details and legal debate. It is a rare report that has its recommendations implemented to the extent that even its commissioners wish. The impact of inquiries "on the processes of policy analysis is generally more significant than their capacity to deliver acceptable recommendations" (Pross, Christie and Yogis 1990: 16). As the secretary to one royal commission explained, the commission's influence extends far beyond the recommendations. A commission exerts a general influence on policy discussion, it pulls together and generates a significant body of research, it provides an intensive education to the personnel of both the commission and the key interested players, and the hearings themselves "tend to pull together the players on a particular issue. They force those players to listen to one another. They force those players, particularly umbrella organizations, chambers of commerce and the like, to pull their own membership together on a particular public policy issue" (Godsoe 1990: 72–73).

"FOR THE PUBLIC"

Inquiries show a questioning public that government is able and willing to establish truth and to manage problems. Reports encapsulate the bureaucratic spirit of their times. By reading them closely, we can see, built into their literary techniques, the kinds of agents who were constructed to act on—and on behalf of—the community, the agents they aimed to reform and the ways in which knowledge and power were presented to the community as working *for* them. This is what commissions of inquiry do: they have as much in common with truth and reconciliation commissions as they do with criminal trials and scientific studies. The report of the *Ocean Ranger* inquiry analyzed policy, but more importantly, it validated ways of knowing that reasserted the legitimacy of governments and companies. In sociological terms, the report was a "reality report" (Green 1993). That is, it clarified and secured a public account of how the men died, and it did so in a way that seemed to reconcile the ways people make sense of things in day-to-day life with the decision-making processes of governments and corporate bureaucracies.

Recalling Justice Hickman's awareness that he and his commissioners and staff wrote this report "for the public," I read it as a work of literature as well as a statement of facts. From witness testimony and material evidence, the commission weaves a tale that combines technical, social scientific and journalistic reporting with everyday storytelling. It is worth reading the report with a view to determining which "public" or imagined community it addresses and what central message it delivers. What was at stake here was nothing less than the translation of the historical narrative into a story that did not blame the oil companies, that cautioned against continued dependence on Ottawa and that turned a cautionary tale into an endorsement of even greater public trust in the promise of oil. The inquiry's main role is to show the legal system effectively gathering evidence and synthesizing it through the cooperation of bureaucrats, scientists and industry experts, and then to write "the" story: the version of events that will from that time forward be the starting point of all other versions.

THE AUTHORITY TO TELL THE STORY

Report One is dramatic reading. The opening sections position the commission as an authoritative source of legal and scientific facts. Where witnesses are accessible, the commission cross-examines their testimony in a publicly broadcast, quasi-judicial hearing. This process secures material facts, the tone of the relations on the rig and an account of the "customary practices" on the rig. Where witnesses are either dead or unwilling to participate, the commission uses scientific expertise to draw the truth from the material traces the men left on evidence retrieved from the ocean floor.

The preface grounds the report's legitimacy on the community's need to know what happened. For example: "In that tightly-knit maritime community there were few who did not discover a link, direct or indirect, to one of those lost in the tragedy. The inquiry by this Royal Commission was therefore of unusually deep concern to Newfoundlanders. It also had important implications for the rest of Canada and for other maritime nations engaged in the search for offshore oil and gas" (iii). To reassure Newfoundlanders and Canadians that their governments had not offered our men up as fodder for profit-driven outsiders, the commission identifies the material cause of the disaster as a mechanical chain of events starting from the broken portlight and even the rig's designers and builders. It identifies the socio-political cause of the disaster as the immaturity of the international oil industry and its regulators. The commission's mandate was to examine "not only the cause of the loss, but also areas of vulnerability within which lay the potential for this disaster and the seeds for future ones" (iv).[2] Thus the report revises the regulatory failure into a scientific mystery: "The investigations of the loss of the *Ocean Ranger* and its crew should go beyond the realm of acceptable conjecture or reasonable deduction based upon circumstantial evidence. It should endeavour through scientific investigation to determine why the *Ocean Ranger* alone of the three rigs on Hibernia, capsized and sank during a severe winter storm" (iv).

Broadcasting the inquiry testimony helped to restore the promise of social development through technology, as a model of new media—"new" in 1982 terms—with the testimony recorded and indexed for posterity. The "public" addressed and, to an extent constructed, by the report is a Newfoundland in Canada and a Canada in the international petroleum business:

> Canadians from all parts of the country are now employed in exploratory drilling operations off Eastern Canada. Responsibility for their safety and for the proper conduct of this major new industry in Canadian coastal waters has been assumed by government both nationally and provincially. The international maritime and oil industries have a keen interest in how these responsibilities are administered. Much has been achieved by governments and the industry over the past two years. But a great deal still remains to be done. (v)

SURVIVAL OF THE FITTEST

As the report tells it, the *Ocean Ranger* and its crew were lost, first and foremost, because the international oil industry grows through a "gradual evolution of offshore technology" (viii). In an epic struggle for survival, oil exploitation crawled off the land, through the swamps and into deeper and deeper waters, harsher and harsher climes. The "growth" of demand, the "drive" for national self-sufficiency and the "surge" of exploration meant an inexorable evolution in practices: "The unique nature of this industrial-marine endeavour, together with the constant evolution of new technology, has presented a challenge to agencies established to set standards and govern the design and activities of more traditional craft" (viii). According to this Darwinian logic, in an environment of vital energies, some die off while others adapt. The metaphors of natural Darwinism in the story of the evolving oil industry also suggest a social Darwinism, according to which the "tightly-knit community" of Newfoundland would need to adapt to the rugged individualism of oil culture. Newfoundland and Canada needed to adapt in two ways: land-based oil culture would have to adapt to the water, and marine-based communities would have to adapt to the profit-drive of the oil industry.

Regulation and corporate capitalism have always co-existed, the report reminds us as it describes the "regulatory structure" that set the stage for the disaster. As early as the late 1700s, in fact, the official ships list from Lloyd's of London let investors know which ships were insurable. In those days, however, regulation protected workers or the environment only insofar as an injury to them meant a loss to shareholders.

In the early days of oil and gas development in our offshore, Canadian and Newfoundland governments left crucial regulation to the American Bureau of Shipping and the American Coast Guard. At every possible level of regulation, the would-be regulator turned a blind eye and the *Ocean Ranger*'s operator and designer, ODECO and its business partners, took the opportunities afforded by the paucity of regulation to neglect even the most basic safety standards set by industry, common sense and respect for human life. For ODECO, regulatory gaps were opportunities to ignore safety, not opportunities to "self-regulate." To cite just one example in the litany: "The lifeboats and life rafts on the *Ocean Ranger* did not meet U.S. Coast Guard requirements. It has also been determined that the rig was not manned according to requirements of the Certificate of Inspection, and that its Cargo Ship Safety Equipment Certificate... had expired" (8). As for Canada's regulatory responsibility as the "Coastal State," "Because of an unresolved jurisdictional dispute over the ownership of offshore resources, each government enforced its own requirements on Grand Banks drilling operations" (8). The report shows but does not comment on the fact that for the disputing governments, "regulation" had meant control of rights to exploit resources. "Representatives of both COGLA [Canadian Oil and Gas Lands Administration] and the [Newfoundland and Labrador] Petroleum Directorate admitted that they did not treat the safety of the rig's marine operations as a priority. Since the sinking of the *Ocean Ranger*, the U.S. Coast Guard, COGLA and the Petroleum Directorate have all increased the rigour of both regulations and enforcement policies" (9).

DISASTROUSLY DESIGNED BALLAST CONTROL AND UNTRAINED WORKERS

In a chapter entitled "The *Ocean Ranger*," the report describes the rig as a ship where men lived and worked, without mentioning drilling. The report recalls the lost possibility of a safer operation, calling attention to the lack of mindfulness about the rig as a workplace, a kind of ship and of the marine environment. It compiles a string of missed warnings and incidents upon incidents of negligence. For example, ODECO complied neither with the U.S. Coast Guard nor with its own policies regarding ballast operations training and support. ODECO also ignored the American Bureau of Shipping's warning that the chain lockers were "the first point of flooding" and should be protected. Nor did ODECO provide cold water immersion suits, even though COGLA "suggested" they be available to all workers, as part of the "lesson learned" from the loss of the icebreaker the Arctic Explorer when thirteen men died but nineteen men survived for two days in safety suits and inflatable rafts (25).

A detailed account of ODECO's systematic neglect of ballast control builds with terrible tragic tension: we know where this is heading but, like the crew that night, we are helpless to interrupt. The ballast control room was in reach of the waves, in the leg of the rig, ostensibly so that the workers could do visual checks out the windows.

> As the ballast control room was considered a dry area, the ballast control console was not protected from sea water. Each porthole did have on the inside a hinged metal cover or deadlight, which could be secured over the portlight to provide protection, but the normal practice was to leave these covers open. Even though the tempered glass was unable to withstand the pressures generated by waves predictable under extreme storm conditions, there was no protection provided for the console in case the portlight did break and sea water entered the room nor was the console itself designed to be watertight. In the event of the accidental flooding by sea water the operation of the ballast control system could be affected. (18)

That fatal little window was a crucial link in the mechanistic chain of causes that led to disaster. Even after the window broke and the ballast control panel was covered with seawater, the men had chance after chance to turn things around. Everything they needed was on hand, but ODECO had neglected to tell them how to use it. "ODECO ought to have realized the importance of providing a method of manually controlling the ballast valves from the ballast control room and incorporated this requirement in its contract specifications" (20).

The links of the fatal chain of events might have been broken by a mindful intervention by ODECO—the rig was equipment, certainly, but it was also living and working quarters, inhabited and ultimately managed by men. "But there were no diagrams or instructions regarding the use of this method of manually controlling the valves from the ballast control room" (20). On the evening of the disaster, once the portlight broke, the solution to the rig's tilt was there in the ballast control room, but it was worse than indecipherable. The unexplained manual control rods were an enticement to intervene in ways that sped the men to their deaths. Nor was there any way for the men to talk to anyone as they tried to make sense of their situation: "Surprisingly, no sound powered telephone was installed in the ballast control room" (22).

Key personnel are developed as characters in the action of the report's reconstruction of the night's events. The Mobil toolpusher (on an oil rig, the manager in charge of drilling), Kent Thompson, came from the U.S. He had extensive experience as an oilman, but no experience with the marine aspect of the operation. The so-called "master" of the rig was a certified sea captain with no experience on oil rigs and no understanding of the ballast control system. His experience in the previous decade was as "a stevedoring superintendent, as a technician in a detoxification centre and as a salesman" (30). The report made a point of specifying the nationality of each of the senior staff people. One was Canadian, Jack Jacobsen. The rest were American citizens.

There was no regulation stipulating that crew be trained as ballast control operators, so ODECO did not train them.[3] "Local hiring policies in Canada, as they were applied to the offshore drilling industry, were complicated at the time of the loss by the existence of a dual regulatory system" (35). Did Newfoundland's drive to create jobs contribute to compromising safety? "There is no evidence that the insistence upon the hiring of local residents caused or contributed in any way to the loss of the rig and its crew" (36). So why devote three pages of the report to this question? The report closed this question authoritatively, though members of the oil industry would raise the issue repeatedly in their response to the disaster.

Dreadful aspects of the evacuation included the confusion and terror of the tilting rig, the screaming winter storm, the waves, the scarcity of safety suits and working lifeboats and the men's knowledge of almost certain violent death. The report manages the reader's anxiety masterfully. Launching the new lifeboats without training would have meant almost certain death: the boat swinging from a davit that did not release; the men slammed against one another, slammed against the sides of the lifeboat, until the vessel itself was finally dashed against the rig.[4] This is one of the most difficult things to think about in relation to the so-called evacuation. What if the men got into those enclosed boats and then were crushed because they did not have a clue how to use them? Two of the four lifeboats were found, badly damaged. A third lifeboat was found and it is believed to have remained right where it was: stored on deck. No trace of a fourth lifeboat was ever found. The report addressed an even more terrible possibility: that men might have taken refuge in an airtight part of the rig only to sink to the bottom of the freezing North Atlantic. This was a possibility entertained even by some of the members of the dive company that retrieved the material evidence for the inquiry. However, everyone agreed that this was unlikely and that hypothermia would have claimed anyone still in the wreck long before their air ran out.

The chapter titled "Operations" draws on the incident of eight days before the disaster, when the rig developed a bad list and the ballast men could not figure out how to correct it. It was so bad that the men were called to muster at the lifeboats. The master was called to the ballast room, but it was clear to everyone, including the toolpusher, that the rig's master had no idea how to operate the ballast system.

> After the list had been rectified, Thompson, the toolpusher, in the presence of Jim Counts, ODECO's shore-based drilling superintendent, severely criticised Captain Hauss for causing the list and told him to be sure that it did not happen again. Captain Hauss agreed, according to Porter, not to operate the ballast control console again. Counts testified that both he and Thompson had lost confidence in Captain Hauss, but neither took any action to replace him immediately, a fact which reinforces the impression that the master was only on board in order to comply with U.S. Coast Guard regulations. (50)

This "prelude" of the week before (as the report called it) demonstrated other things, too: COGLA regulations called for operators to develop a "plan" for emergencies. But there "was no evidence to indicate that ODECO personnel either on shore or on the rig were familiar with its contents" (48).

Chapter Five, "Events Before Evacuation," which builds from the early weather reports into a hurricane-level winter storm, documents the grave miscommunications among the men on the rig, their failure to understand the situation, and the crisis and speed of the final hour. At 1:08 a.m. the first mayday came from the rig, and at 1:30 the last contact with Ken Blackmore informed that the men were taking to the lifeboats.

Part of the problem in the final hours was that Mobil's toolpusher did not seem to like or respect Jimmy Counts, his main resource on the shore. The men in the ballast room misunderstood just about everything. The report quotes the now dead men, their echoes overheard by the radio officers on neighbouring rigs. "According to King, Dyke also said that 'they were getting shocks off other equipment... [and that] valves were opening and closing on their own'" (61).

MAKING SENSE OF THE WRECKAGE

Silence follows the 1:30 radio transmission that the men were at the muster stations and preparing to evacuate. At this crucial moment, when the reader is empathetically engaged with those terrified men, the report turns to the technological challenge of developing offshore oil production. With the men silenced forever, the report seeks the traces of their deliberations etched in the wreckage. The National Research Council built a scale model of the rig and tested it in a huge basin. The pontoons of the *Ocean Ranger* were 400 feet long, the water was 250-odd feet deep. The engineers were unable to reproduce a small-scale re-enactment of the precise way the rig's interior took on water once it started to go; how the water rushed into the chain lockers then down vents and into the living quarters, storage and other spaces below. To capsize that rig it was necessary that the men be not simply untrained, but that they be trained to do exactly the wrong thing. Their training told them that turning the mysterious rods closed the valves in the pontoons when in fact it opened them (77).

Various experts deciphered the traces the men's final efforts left on the wreckage. ODECO's lawyer is more present in this chapter, as he tries to destabilize the causal chain leading from the deaths back to the decisions to keep drilling even as the neighbouring rigs stopped. He raises the possibility that debris rather than inadequate glass strength burst the porthole and that individual worker ineptitude started the "fatal chain of events." A careful reading of this report drives home the extent to which the *Ocean Ranger* disaster was the result of design flaws that expressed a lack of mindfulness about the rig's environment.[5]

The material evidence speaks for the men, who, it seems, turned the power back on to the ballast control panel some time between 12:30 and 12:45 a.m. Why they did this was a mystery. Noting this, the report shies away from the men's thoughts:

> Whatever may have been the reason behind the action, the restoration of power allowed random microswitch short circuits to open the correspondingly remotely operated valves. It is known that the rig incurred a sudden port bow list, and it is concluded that the cause of this list was an ingress of water from the sea into the port pontoon. (94–95)

Like the vital energies driving the evolution of the oil industry, the water became an active agent:

> It is also likely that water flooded from the lower deck into the column spaces above the chain lockers, into the accommodation areas through damaged portions of the superstructure and thence into the lower deck area, and into the lower hull tanks through exposed vent lines. (95)

The report refuses to speculate whether the flooding began while men were still below: "Whether or not any of this damage and flooding occurred before the decision to abandon is unknown" (95). ODECO's counsel pushes towards an even stronger conclusion of worker error, but the report resists:

> It is difficult to accept the argument by ODECO counsel that the ballast control operator would insert the manual control rods as a precautionary measure if the ballast control console was operating normally. The operators on board had never used the rods and accordingly it is highly unlikely that they would have used them particularly when, as

> suggested by ODECO, deballasting had been completed and no further ballasting operation was required or anticipated. The insertion of the rods as a precautionary measure suggests a degree of planning and thought that is not consistent with the manner in which the rods were actually used. (96)

The use of the manual rods to try to override the malfunctioning electrical panel was a "hasty rather than a planned activity" (96).

Finally, the report casts the "crew initiated" sinking of the rig as originating in failures of design and management:

> Each event and action which contributed to the loss of the *Ocean Ranger* was either the result of design deficiencies or was crew-initiated. The disaster could have been avoided by relatively minor modifications to the design of the rig and its systems and it should, in any event, have been prevented by competent and informed action by those on board. Because of inadequate training and lack of manuals and technical information, the crew failed to interrupt the fatal chain of events which led to the eventual loss of the *Ocean Ranger*. It is, nevertheless, the essence of good design to reduce the possibility of human error and of good management to ensure that employees receive training adequate to their responsibilities. (99)

The report concludes its use of material evidence and peeks into the darkness and silence of that final hour and a half or so of life. It lists "contributing factors" as a causal chain, starting with the "design decision to locate the ballast control room in the third starboard column below the lower deck" and the failure to use strong enough glass in the portlights (100). In this way, the report acknowledges the men's roles in the chain of events —thus implicating the government of Newfoundland and its local hiring policy—but ultimately places responsibility on ODECO's shoulders: "The individual factors contributing to the loss had lain dormant on the rig throughout its working life, but it was their unique active combination on February 14–15, 1982, which caused the tragic event" (99). The men's desperate final attempts to recover from the list were described as "incomprehensible."

UNTRAINED EVACUATION AND RESCUE ATTEMPTS

The report then picks up the story of the men's "evacuation." In the chapter titled "Evacuation and Emergency Response," we turn back to the men, though we no longer seek access to their motives and decisions. The report leaves their last moments to them. As well, explicit speculation about how the men who launched Lifeboat No. 1 died is avoided. The report admitted one cause of death and only one, based on the autopsies of the recovered corpses: "In all cases the cause of death was drowning while in a hypothermic condition" (115). The last words from the *Ocean Ranger* are repeated with the addition of Jacobsen's saying to the *Sedco 706* that the rig was "'listing... and not coming back for us so we need every helicopter in the air we can get out here....' His voice during this conversation was surprisingly calm" (106).

The *Seaforth Highlander* crew sped back towards the rig, the "visibility of the bridge was limited by heavy seas and blowing snow," until finally they saw the rig, fully lit. "As she moved closer, clusters of white lights and smoke flares were visible off the port beam. Upon inspection it was determined that these lights were attached to life preservers floating on the water. The life preservers were empty" (108). Life preservers with no life. Then, amazingly, there was a distress flare. Then another.

> At 2:21 a.m. the *Seaforth Highlander* reported to the *Sedco 706* that it had spotted a lifeboat and was proceeding toward it. This information was immediately passed on to Mobil's shore base and to SAREC [the Search and Rescue Emergency Centre]. Graham testified that he issued instructions to Fraser to advise the masters of the supply vessels not to secure lines to lifeboats. In his testimony Graham explained that he was aware of an incident in the Gulf of Mexico in which a lifeboat had capsized while under tow. Fraser stated that the instructions were relayed to the supply vessels, but both Higdon and Duncan testified that they did not receive these instructions; nor did the other supply boats, the Boltentor and the Nordertor, have any record of receiving them. (108)

If I am reading the muster list correctly, and the evacuation was organized in any way, then my brother Jim could well have been in this Lifeboat No. 2. Better this, I think, than for him to have been on Lifeboat No. 1, which was probably smashed when the men tried to launch it and get clear of the sinking rig. The *Seaforth Highlander*'s crew described the men in the lifeboat who "could be seen moving about. Some of the men were bailing" (110). We do not know who these men were and we do not have access to their perspectives.

> Woolridge then threw a line, with a life ring attached, to a man in the aft hatch of the lifeboat. The man caught the line and made it fast to a handrail on the canopy of the lifeboat; Jorgensen tied the other end of the line to the crashrail on the port side of the *Seaforth Highlander*. Meanwhile, seaman Rees threw a second line with a life ring attached;

> this line was made fast to the lifeboat by a man who appeared from the bow hatch and to the crashrail of the *Seaforth Highlander*. (110)

Nameless men emerged from the lifeboat.

> These men were wearing hard hats and either work vests or life preservers; some were lightly clad while others wore heavier clothing. The lifeboat began to roll slowly to port, away from the *Seaforth Highlander*, and within seconds capsized throwing the men who had been standing on the port gunwale into the sea and snapping the lines which had been attached to the *Seaforth Highlander*. As the men from the lifeboat spilled into the sea, the water in the immediate area was illuminated by the lights attached to the life preservers. The lifeboat had completely capsized. (110)

The report gathers itself with an officious, "The time was 2:38 a.m."

The Mobil shore manager knew how not to kill these men. However, because oilman culture denied that the *Ocean Ranger* was a ship and not just an oddly located drilling operation, he did not effectively communicate this to the men on the supply ships, who were, themselves, frozen, battered by the storm and traumatized by their failure: "In spite of the hazardous and difficult conditions on the afterdeck, Jorgensen narrowly missed grasping a man who was washed against the port side of the supply vessel" (110).

Finally, the last spark of life from the *Ocean Ranger*: "One or two of the men in the water were able to hold onto the capsized lifeboat longer than the others" (110).

THE SOCIO-POLITICAL WORK OF THE REPORT

The report narrates the evacuation of the rig and the rescue attempts from the perspective of the living witnesses: the men on shore, the radio officers and ballast operators of the nearby *Sedco 706*, the crews of the supply ships and ultimately the Search and Rescue crews. The report does not invite us to imagine what those men experienced in the lifeboats and the water. The shifts in perspectives in retelling this story matter in many ways, potentially even for the financial settlements negotiated by families at the same time as the inquiry's hearings were being televised. If the families gained access to U.S. courts, they might be compensated for punitive damages against the negligent companies and for the men's pain and suffering, as well as for the usual economic loss. "Drowned in a hypothermic state" was one thing, "crushed, while trapped in an enclosed, vomit-filled space, suspended by a malfunctioning, sub-standard release mechanism over battering waves" was quite another.

The first report of the Royal Commission of inquiry gave shape to disruptive questions. How did the promise of oil turn into tragedy before investment in that promise even started to pay out? How did those companies get into Canadian waters, with Canadian crews, without reasonable safety equipment and protocols? How could the oil industry operate off Newfoundland with such obvious contempt for, and ignorance of, maritime culture? Who decided to leave the ballast control room in the leg of the rig while it was placed on the deck in the *Ocean Ranger*'s otherwise identical sister rig? How were relations between onshore and offshore, and between the men in the ballast control room and the men calling the shots, so distant that the ballast control men had no real support as they struggled to respond to the worsening tilt and the men calling the shots had virtually no advance notice that the rig was in imminent danger? What was the policy behind pushing oil companies into local hiring without insisting also on local training programs? How did a rig whose pontoons measured 400 feet capsize in water that was only 250-odd deep? How and why did it sink so fast?[6] How could the rescue attempt be so disorganized and ill-equipped, coming less than two years after the loss of the *Alexander Kielland* and its 123-man crew?

The answer to all of these questions is obvious: Canada and Newfoundland behaved like U.S. colonies, leaving safety regulation to foreign authorities, while the oil companies took freedom from regulation as an opportunity to ignore safety and not as a chance to self-regulate. An important role of the inquiry was to manage a potential crisis in confidence in governments, and so this obvious answer needed to be revised into a more complex set of technical and regulatory issues. The televised inquiry proceedings reassured a skeptical public that by working together, the governments of Canada and Newfoundland had the power to authorize truth by making the companies answer to the public. The community as a whole revisited the details to the point of exhaustion through the inquiry. The relentless parade of bad news resolved itself into fatigue: "As the hearing dragged on, giving it the unenviable record of being the longest of all Canadian Royal Commission inquiries, pressure was on for an end to it" ("Here and Now" August 10, 1984). Even the number of family members attending the hearings dwindled.

As a reporter from Canadian Press described the last day of public hearings: "It was almost like the last meeting of a club about to fold" (*Cape Breton Post* March 13, 1984). The one outsider to this "relaxed" camaraderie was Cle Newhook from the Ocean Rangers Families Foundation, who noted the absence of the families' lawyers. Now that the settlements were paid, he said: "It seems they just wanted to take the money and run" (*Cape Breton Post* March 13, 1984).

Now that the public had their fixed chain of material cause, leading through the portlight back to the design flaw, the report could add the men's involvement. As the report explains, in what is probably the most quoted passage from the report: "Despite the failure of the portlight and the malfunctioning of the ballast control panel, the loss could have been prevented by knowledgeable intervention on the part of the crew. Indeed, had the crew only closed the deadlights, shut off the electrical and air supplies to the panel, cleaned up the water and glass and then retired for the evening, the *Ocean Ranger* and its crew would have survived the storm that night" (139).

The public agenda gradually turned to the future: what was to be done? The Newfoundland public watched, listened, discussed and the grew weary of the topic and leery of the expense: Hickman's imagined reader, "the public," were actively engaged in writing the story themselves. The report then presented the *Ocean Ranger* story in a form that made such clear sense to its audience that it was as if "the public" had written it themselves.

Few people read the report themselves. Instead, most received a version from professional storytellers: lawyers, journalists, politicians and industry leaders. This public story constructed a causal narrative with clear recommendations for reform so that this "will never happen again." After eighty-eight days of public hearings and between thirteen and fourteen million dollars spent on the inquiry, "No one is expecting much of a surprise in the Commission's conclusions about why the rig sank. The chain of events which started with a porthole window breaking and concluded with a failure of the rig's ballast and pumping system to correct this problem is well known by now. What will be of greatest interest will be the Royal Commission's recommendations on what's to be done to prevent such a thing from happening again" ("Here and Now" August 19, 1984). The first report replaced the most divisive facts of personality and cultural conflict, poor regulation, wrong-headed design and general complacency with the spectacle of the federal and provincial governments working together to wrest truth from the oil industry.

REPORT TWO AND INDUSTRY RESPONSES

The first point in the "lesson" of the *Ocean Ranger* disaster was the apparent surprise that the North Atlantic is a dangerous place to drill for oil. Having discovered this shocking fact, Newfoundland, Canada and the oil industry put their heads together to come up with a scheme to make it as safe as possible.

Sarcasm aside, Hickman seems to have seen that relations between bureaucrats, industry leaders and technical scientists had to be reformed. They needed to come together to exchange ideas apart from the bombast of federal-provincial relations. Hickman paused, in a public address, between the two phases of the inquiry:

> The [phase] that began in March and is now concluded, has dealt with the sequence of events surrounding the loss. This evidence has taken the form of the, at times, dramatic and extremely moving accounts of those who were involved in the tragedy. We have deeply regretted having had to focus attention once more on these painful events. This evidence, though, is crucial because it is the only human testimony relating to the loss by people who were there that is available to us. (Hickman 1983)

The forensic phase being over, it was time to write the story and turn to the future. Hickman continued: "The restoration of confidence is a key purpose for us all." Pressure was mounting to conclude the lengthy process and to limit further spending—at a rate of six million dollars per year, shared equally between federal and provincial governments—on an inquiry that had already convinced the public that it knew what happened and largely who was to blame.

The Royal Commission convened a conference: "Regulation in the Offshore." The obvious irony is that this conference should, by rights, have taken place long before February 14–15, 1982. Leading thinkers in the oil industry, maritime law and risky enterprises like atomic energy were invited to Memorial University, where they spent three days presenting papers and debating. Their focus was not on "what happened" but on trying to untangle the web of responsibilities, decision-making and technical capabilities to make the hazardous project of drilling in the North Atlantic as reasonable as was affordable. As one journalist described the event:

> To decide what constitutes acceptable risk on the drilling platforms of the Northwest Atlantic, the Commissioners had to wade through a complex set of discussions, descriptions, and arguments that touched on practically all aspects of offshore drilling. They were told of the sea climate, and how rigs are designed to withstand the worst-case scenario of the "hundred-year storm." They heard how workers could be saved from rolling drill rigs in emergencies, and how people interact with machines and one another in normal times and crises. Finally, they listened to a long, heated discussion on the best means of regulating the industry to ensure that safety procedures are followed. (Campbell 1984: 9)

In the debates seen by this journalist, the "sharpest" criticisms were levelled by the president of ODECO engineering—the designers of the flawed rig—at the classification societies that are responsible for regulating rig design and construction. The classification societies, he argued, had gradually eroded the safety margins that govern rig stability because they permitted a designer to assume that "all 'damage events' will occur only in calm seas and in winds never exceeding 50 knots, they also allow the rig deck to become submerged with seawater—'knowing full well that the deck isn't designed to withstand sea wave forces driven by winds as low as 50 knots'" (Campbell 1984: 15).

In the second and much less read report, the commissioners professionalized the responsibility for offshore safety, suggesting that it was a natural outcome of healthy corporate life and not a regulatory responsibility of governments. The *Ocean Ranger* disaster was understood as a communications failure between the various agents ethically responsible for the rig's safe operation.

> A polarization certainly showed up early in the conference between those who favour more government intervention in the oil industry and oil companies who oppose this. Those favouring more government intervention think there should be stricter safety rules for the oil companies to follow. The oil company representatives prefer self-regulation and say they can do a good job of ensuring safe practices offshore. ("Here and Now" August 23, 1984)

W.G. Carson, author of *The Other Price of Britain's Oil*, grabbed attention by speaking so forcefully about the need to regulate for safety that oil executives "jumped to their feet in the conference." The second report lists three categories of regulatory failure: lack of monitoring and enforcement; regulatory deficiencies pertaining to operational procedures, training and qualification, rig stability, emergency preparedness, design and construction; and "the jurisdictional complexity and ambiguity" of Newfoundland's offshore (Hart 2005: 72). By the release of the second report, the commission had adopted a "decidedly conciliatory and dependent attitude," as one commentator notes, citing this example: "It is generally recognised that there are certain areas such as industrial training where industry will know best what standard is necessary for competent performance or where standards determined by external agencies are the best to be adopted" (in Hart 2005: 73).

The oil industry responded to the inquiry's findings of shoddy operations, design flaws and the need for improved regulation with a counter-version. They repeated their central point: far from being the key to the future, regulation was the curse of the past. In their version, regulation, not corporate negligence, sank the rig and killed those men. For them, plain bad luck and the audacious risk of entrepreneurial initiative, in combination with local hiring regulations, created the disaster. The lesson for them was that governments needed to learn to let oil companies do their jobs. As reported in the conference proceedings from an industry symposium, "Safety of Life Offshore" in March 1983, "the capability and judgment of the experienced rig designer is more and more being replaced by arbitrary dictatorship of bureaucrats in many countries" (Bennet and Goldman 1983: 2).

The oil industry—rig designers in particular—presented itself as a creative community driven by competition in harsh environments. The real danger signalled by the disaster was that if industry does not take initiative to improve safety then the natural evolutionary creativity of oil technology will be stunted by bureaucratic regulation. These oilmen worked in the spirit of nineteenth-century entrepreneurship, hand-in-hand with the insurance industry. As an English underwriter explained, "safety is... a comparative and qualitative subject difficult to visualise and assess. Safety and risk are complementary and opposite terms—most safety least risk, least safety most risk—and the graduations of juxtaposition are infinite; or in another way—safety and risk are in an inverse proportion to each other" (Tucker 1983: 78).

Given the level of negligence in ODECO's rig design and their erroneous training of the ballast operators and ineffectual sea master, it is remarkable that ODECO recovered insurance money. If only my car insurance would cover me if I had an accident and my passengers died when my seat belts were not working, my car was not inspected and my licence to drive expired.

The underwriter Tucker gave a brief history of insuring the offshore oil and gas industry, arguing that the relationship experienced a crisis in the mid 1960s, when insurance coverage skyrocketed following catastrophic losses. "In the spring of 1966 the London Drilling Rig Committee declared a moratorium on all new underwriting or even renewals of existing policies.... Two vital subjects were under discussion: one, the very existence of the market and two, the price upon which the risk of mobile offshore drilling units could be accepted" (Tucker 1983: 79). The insurer rejected the industry's attempts to narrow the "cause" of accidents to technical chains of bad luck and equipment failure and invoked a commonly quoted ruling:

> To treat *proxima causa* as the cause which is nearest in time is out of the question. Causes are spoken of as if they were distinct from one another as beads in a row or links in a chain, but if this metaphysical topic has to be referred to—it is not wholly so. The chain of causation is a handy expression but the figure is inadequate. Causation is not a chain, but a net. At each joint, influences, forces, events and precedent simultaneously meet: and the radiation from each joint extends infinitely. At the joint where these various influences meet it is for the judgment as upon a matter of fact to declare which one of the causes thus joined at the joint of the effect was the proximate and which was the remote cause. (Shaw, in Tucker 1983: 79–80)

Coerced into metaphysics, Tucker explained that a "net" of causation was more adequate than the image of a chain of technical mishaps. He then mentioned the *Ocean Ranger* along with the *Alexander Kielland*. This underwriter pondered: "Man constantly strives to understand and control his environment. However, as each technological advance is made so he opened the door to an endless set of questions yet to be answered. Some of these questions raised the problem of certainty of loss" (80).

Insurance providers saw themselves as responsible for paying for "accidents" or "a fortuitous event" which "is dependent on chance," but not where damage was inevitable. The underwriter went on to mention the *Ocean Ranger* in damning detail but this time without naming the rig or ODECO:

> Recently, a drilling vessel was operating in its normal and intended fashion, but for the reasons stated in the extract, capsized with considerable loss of life, and the subsequent accusations have been made:
>
> Failure to provide a reasonably safe place to work
> Failure to exercise legal duty to reasonably care for passengers or crew
> Failure to provide adequate measures for the maintenance and safety inspections of the vessel
> Failure to follow procedures for the compliance with maintenance and safety measures required by laws of the United States especially as implemented and applied by the United States Coast Guard
> Failure to institute and implement safety measures necessary to create and provide a reasonably safe place to work
> Failure to provide adequate personnel and adequate procedures necessary to create a safe working environment
> Failure to create and operate a seaworthy vessel able to withstand the conditions appropriate to the vessel class
> Failure to oversee and supervise, and to consistently institute and apply procedures and practices through servants, agents, or employees, to provide reasonably safe operation of the vessel
> Failure to equip and maintain the vessel with adequate lifesaving procedures necessary for a successful and reasonably safe evacuation of the vessel, or to provide an adequate, trained, skilled crew to accomplish the same, numerous other failures to act, or acts of negligence
> Failure to observe an absolute and nondelegable duty to provide a seaworthy vessel and seaworthy crew, and to conduct the operation of the vessel in a seaworthy manner. (80)

The underwriter sides with the oil industry against government regulation, but he extols the new British move to sign an agreement with the oil industry to "self-regulate." The real stick driving reform may be expressed here when the underwriter, and indeed the insurance industry, threatens to withdraw from insuring mobile offshore drilling units altogether. Despite the insurance industry's zeal for facilitating the extension of man's control over nature, at a certain point the industry made it clear it would have to cut its losses. The underwriter explained that the insurance industry lost money on offshore drilling and did not necessarily make enough from other aspects of oil and gas industry to make up for that. The, at that point, excellent safety record of oil exploitation in the North Sea skewed the rest of the world market (Tucker 1983: 82). Tucker concluded: "Philosophically, one can accept that the responsibility of insurers is to ensure that commerce can operate, and that the principle of 'the loss falling lightly on many rather than heavily on few,' is still sound. However, as I said earlier, the stockholder pays the piper and we must heed the tune" (83).

"THE" *OCEAN RANGER* STORY

When I asked Justice Hickman what, for him, thinking back from an almost thirty-year vantage point, were the most remarkable moments of testimony in the inquiry, he did not hesitate. He seemed genuinely angry when he described the testimony from a ballast control worker from America that illuminated how completely ODECO had failed to meet even its own modest standards for training on the *Ocean Ranger*. He choked up when he repeated the story of the matter-of-fact heroism of the supply vessels' crewmen and the frustration of learning that "a Mobil Oil employee" failed to effectively communicate to would-be rescuers: "Don't put a line on it" because the lifeboat would capsize. Justice Hickman recalled resistance from the companies, though he spoke with warmth and professional admiration for ODECO's lawyers. Mobil agreed to have their foreman and other employees testify but only if it was not televised. The biggest lingering question for Justice Hickman arose during the commission's tour of the *Dyvi Delta*, the sister and exact replica of the *Ocean Ranger*. The identical rigs had one crucial difference: the *Ocean Ranger* kept the ballast control in its leg, following the original design by ODECO, whereas the *Dyvi Delta* had the ballast room more safely located on the deck. "Whose decision was it to put ballast control on the deck?" Hickman asked. "'Talk to the owners,' I was told. But I never did get an answer" (Hickman 2008).

Reassuring the public that Newfoundland and Canada had control over the material evidence was essential to the inquiry's own legitimacy. Mobil provided the vessel that first located the rig in the days immediately following the disaster, and this raised concerns that Mobil had unreasonable access to the evidence (*Evening Telegram* 1982u; O'Neill n.d.).

Ultimately, Justice Hickman was proud of securing the material evidence for the Newfoundland and Canadian publics. Max Ruelokke, then owner and president of Hydrospace Marine, who lost five staff on the rig—including Perry Morrison—advised that diving around the *Ocean Ranger* wreck would be dangerous. Ruelokke's crew recovered the ballast control panel and brought it to St John's. Hickman laughed when he recalled receiving a call from Deputy Minister Peter Troop, advising that the American embassy had contacted Canadian External Affairs. The *Ocean Ranger* was an American registered ship. "There was going to be trouble," Hickman recalled: "'Chief Justice, you are about to cause a diplomatic incident,'" he mimicked, laughing. Despite pressure from ODECO and the American Coast Guard, Hickman had all evidence placed under guard by Canadian authorities and, on the opening day of the public hearings, he denied ODECO counsel's request to restrict the commissioners from considering questions of the rig's structure and design. As the inquiry hearings began, ODECO resisted having its senior employees testify, arguing that they might "reveal trade secrets." This gave the commission a chance to demonstrate its ability to pressure ODECO into having its senior employees testify, even though it was based in the U.S. and the inquiry's subpoena power extended to Canada alone (*Evening Telegram* 1982j).

Most families had accepted financial settlements in December 1983, eight months before the first volume of the inquiry report was published. Margaret Blackmore, whose husband Ken died in the disaster, responded to the report as chair of the Ocean Ranger Families Foundation. She said: "The frustration and helplessness we have always felt has now been taken over by our indescribable anger and a sense of betrayal. The long litany of neglect spelled out in the report is unbelievable and disgusting, the anger we feel towards the companies defies description and we feel betrayed by both the federal and provincial governments. We condemn them for all their neglect" (*Montreal Gazette* 1984). Despite the lingering sense of injustice so keenly felt by family members, the inquiry process had done its job. From the archbishop's call from the pulpit to hold a joint federal and provincial inquiry, to Justice Hickman's use of the inquiry to showcase Canada and Newfoundland forcing the oil industry to rethink their approach to drilling on the frontiers of technology, the *Ocean Ranger* Royal Commission of inquiry was exemplary as a process that revised the collective trauma into a masterful text. In the East Coast Petroleum Operators' Association, Executive Summary: Offshore Safety Task Force Report, October 30, 1983, the *Ocean Ranger* was not mentioned until the "Conclusions and Recommendations" section, where it reinforces the industry line:

> As a general conclusion, the Task Force study determined that the offshore operators and contractors were meeting, and in many cases exceeding, safety regulatory requirements. Survival and evacuation systems on the drilling units were the most appropriate available, however the importance of further development of immersion suits, survival training and personnel sea rescue equipment and techniques were not totally realized until the *Ocean Ranger* sinking. To enable offshore drilling operations to be conducted in the safest manner experienced, competently trained personnel are required. This requirement must not be sacrificed to meet outside imposed objectives. (13)

The industry had not "realized" the safety challenges of winter drilling in the North Atlantic, and they were hampered by "outside imposed objectives."

The recommendations of the Royal Commission included a requirement for oil companies to train the people they hired to keep a rig afloat, to design rigs with a mind to their real working environments, to train everyone in how to work the safety equipment, to provide rescue equipment to the rescue vessels and to provide cold water immersion suits (which a year before the *Ocean Ranger* loss, allowed nineteen men to float for fifty-two hours before being saved off Newfoundland's northern coast). The majority of the report's recommendations were adopted by 1992, with the notable exception of the industry investing in evacuation technology and the stationing of Search and Rescue helicopters in St John's (Crosbie 2009; Hart 2005: 75).[7] It took between four and five years and what Hickman recalls as the tenacious shepherding of the recommendations by John Crosbie through the labyrinth of the federal regulatory bureaucracy for changes to be implemented. Norway acted more immediately in the wake of the *Alexander Kielland* loss, though the consensus seemed to be that the heavy regulatory regime in Norway bureaucratized the workplace to the point where unions had trouble convincing workers not simply to rely on the rules but to actively protect themselves (Hickman 2008).

In terms of reconstituting the promise of oil and working through the public trauma of the loss, the recommendations marked a distinct turning point. The time for visiting the ghosts, for seeking blame and causation, was over. It was time to turn to the future. While appeasing the families' need for public recognition and change, the recommendations also provided a checklist for reform, thus reinforcing the learning story.

If we recall the promise that Newfoundland could "grow up" through participation in the oil industry, we can see how this was revised into a broader story about how the promise of oil entailed a long-term, international evolution of know-how. The *Ocean Ranger* disaster became one episode in a broad narrative of human progress, situated in the story of the "Newfoundland people," that is, a people unified by "the sense of being a frontier, of being on the edge of things, along with the sea and antique culture" (Earle 1998: 94). The inquiry report stabilized an authoritative version of what happened, who was to blame and what should be done. In the process, it demonstrated the state's power to produce "the" authoritative version of events. The inquiry and its report limited the potential legitimation crisis triggered by the disaster by defining the *Ocean Ranger* story as a "learning story," with Newfoundland, Canada and the international oil industry cast as vigorous agents poised for a promising future of harnessing science and investment for a maturing people. The "lessons" of *Alexander Kielland* and *Ocean Ranger* would not be learned in time to prevent the loss of the *Piper Alpha*, which blew out, burning and asphyxiating 167 men to death within sight of television cameras in 1988. The *Piper Alpha* disaster would be a turning point in attentiveness to safety in the offshore because it was such a hideous—and visible—end for those men.

NOTES

1. "*Ottawa Citizen* 1983. Stephen Neary called for criminal charges once the NTSB [National Transportation Safety Board) report was released. Meanwhile, NDP MP Rod Murphy confronted federal Labour Minister Charles Caccia: "You're responsible. You're a bloody murderer" for delays in occupational health and safety regulations. John Crosbie, then MP for St John's West, accused federal regulators of negligence.
2. The use of "seeds" here is interesting in light of the prevalence of farm imagery in Newfoundland *marchen*. When the stock character Jack heads out for adventure where he will learn to respect the advice of his elders and rely on his own wiles in dealing with dangerous strangers, he often runs into manipulative farmers (Lovelace 2001).
3. "ODECO generally selected its ballast control operators from the drilling crew. If an individual showed the necessary interest and potential, he could train to become a ballast control operator. The stated training program of ODECO permitted a roustabout to train as a ballast control operator after 80 weeks' experience on the rig. After 24 weeks' training he could be placed in charge of the ballast control room. In practice, however, ODECO did not follow this policy. Three former ballast control operators gave evidence at the public hearings (Frank Jennings, Cliff Himes, Bruce Porter). Jennings testified that he responded to a newspaper advertisement and was appointed as a ballast control operator without any drilling or marine experience. After only several days of orientation he stood a normal 12-hour watch by himself in the ballast control room. Himes had 28 weeks, Porter had 32 weeks, Rathbun had 12 weeks, and Dyke had 40 weeks of roustabout experience before being appointed to the ballast control room" (32).
4. "When the *Ocean Ranger* was issued the Certificate of Inspection in 1979, the U.S. Coast Guard directed ODECO to ensure that the lifesaving equipment met

their standards by replacing existing lifeboats and davits with U.S. Coast Guard approved equipment and by installing for 100% of the rig's crew davit-launched life rafts or an acceptable substitute.... At the time of the loss, however, although one of the new lifeboats was installed, it is not known whether it was provisioned and fully operable, and the other was stored on deck awaiting installation. ODECO had not replaced or changed the existing lifeboats and davits to comply with U.S. Coast Guard requirements" (23).

5. "It is concluded that water did in fact enter a number of microswitches in the mimic panel. Once water enters a microswitch, it is virtually impossible to remove the water without physically taking the microswitch out of the switch housing and subjecting it to prolonged drying. The evidence is clear that the microswitches were not removed from the switch housing nor disassembled in any way and that any water that entered the microswitches was still present at the time of the loss" (92).
6. At 12:45 a.m. the men are believed to have turned on the power to the ballast control panel; at 3:38 a.m. the rig disappeared from radar.
7. The government's response to the commission's recommendation indicated that concerned cabinet ministers would "consider the question of federally-funded helicopters and report back to Cabinet" (Canada1986a).

2

THE FAMILIES' FINANCIAL SETTLEMENTS: THE COSTS OF CLOSURE

"A dishonour to your brother's memory"

While the inquiry testimony aired the litany of neglect by governments and companies, the lawyers for the oil and insurance companies negotiated privately with lawyers for the families. The resulting financial settlements close off the families' claims to lawsuits against the companies. Most Newfoundland families settled about half way through the inquiry hearings (Hickman 2008). We were the first family to accept the oil companies' financial offer, as far as I know. Jim was young, single and without dependents, so his case was relatively simple. Other *Ocean Ranger* families held out for as long as four years, first as part of a claimants committee in Newfoundland and then as a few rogue individuals who tried to win the right to sue the parent companies in their home country, the United States. In this chapter, I tell the story of some Canadian families' out-of-court settlements with the companies. I argue that this money has deep cultural importance and plays a key role in restoring public confidence, or at least complacency, about governments and businesses in the wake of a community-rattling betrayal of public trust.

I do not know how the companies split the costs of these settlements or even if having them paid by insurance had any effect on their premiums. The settlements were paid by ODECO, Mobil, Schlumberger, Mitsubishi and others to individual families, some of whom had household earnings of less than $30,000 a year. In considering the sums paid to families to settle the legal threat, it is important to keep in mind the sheer size of the corporations and the amounts of money they spent in daily operations: Mobil paid $95,000 a day to have ODECO staff maintain and operate the rig (Zierler and Bill 1982: 3).

Money's role in the disaster's aftermath is fascinating because while it seems to express what a thing is worth, its social role is much more complicated. Settlement money is highly symbolic for families, for the corporations and for the community at large. As we will see, the way settlements are explained to families changes as the negotiations unfold. At first the lawyers say, "Nothing can replace your loss," then "Here is the equation we have developed to compensate you," and finally, for those who refuse to settle, "You need to accept this offer or they will just make your life miserable."

From the moment family members started to suspect that the oil companies were using information about whether a man was alive or dead strategically, they were, in a sense, in a legal mode. Deliberate delays in communicating that world-transforming news convinced family members that their human dignity was of no importance to the companies. My family was lucky in this limited way; we were spared that indignity and my parents still speak fondly of the young manager from Schlumberger who telephoned them through that terrible time. The use of the news of death for strategic purposes was, for other families, a collapse in basic civility. The work of the aftermath, as we will see, was to bring grieving families back into at least a perceived civil relationship of recognition and interaction with the oil companies and with the community generally.

Money was an issue that was thrust upon families from the very first moments of the disaster. Some family members recall company representatives arriving at funerals with cheques hoping to score an early, small settlement (Campbell and Dodd 1993). American lawyers flew in—some in private jets—from New York, Texas and New Orleans. Two days after the rig disappeared from radar CBC's *Here and Now* reported that if families accepted Workers' Compensation they would not be able to sue the companies. The provincial justice minister cautioned next of kin "not to enter into any agreement without securing professional advice from persons resident in the province" (*Evening Telegram* 1982a), and once again Archbishop Penny used the homily of the memorial service for the *Ocean Ranger* dead to warn families against potentially unethical advances from lawyers.

Payments to the families were crucial to the community's recovery from the collective aspects of the traumatic loss. All the cash—but especially the financial settlements from threatened lawsuits—furthered the community's healing, though at great emotional cost to family members, who experienced the evaluations of their man in dollar terms as a forced, false closure. To accept payment was to return to normal relations within the economic valuations of life in liberal capitalism, and it closed off a time of special status for family members as community members who had been exiled, in a sense, by injustice.

Ocean Ranger family members were offered four kinds of payment, and their role as victims was altered by public awareness of each kind of money, whether they received it or not. Charity, insurance, workers' compensation and out-of-court settlements each played a crucial role in restoring public confidence in the promise of oil. The private insurance payout was distasteful but straightforward. It mechanically assigned a value: it read a chart. Accepting insurance money had at least an obvious contractual logic about it that extended back to the premium paid by the victim. Newfoundland and Labrador's *Workers' Compensation Act* was modified in the wake of the disaster so that eligible families could receive benefits without losing their right to sue. This support came from the province of Newfoundland as a whole.

The methods, purposes and outcomes of the law of tort were far less clear. Tort means "wrong," in the multiple senses of harm, error, contrary to a right and damage. The right to sue is based on a longstanding sense that courts can determine how a victim should be repaid for a wrong done by another individual. Tort law aims "to provide justice to victims and peace and security to society. It serves to punish wrongdoers, deter wrongdoing, compensate victims, and vindicate rights" (Klar 2003: 9). All the *Ocean Ranger* families accepted financial settlements from the oil companies. The various simultaneous legal processes manage incommensurable claims: families think in terms of justice while companies consider the same problem in terms of costs and benefits. Out-of-court settlements close off families' right to sue without the companies admitting any guilt. For the companies, the settlements cost less than hiring lawyers and public relations firms to keep shareholders confident during a well-publicized legal battle with grieving families. For families, the money meant closure, and it changed their public roles. The victims of the early days of the tragedy were recast in different public roles once the settlement money changed hands. "Blood money" taints everyone who touches it, as if the blood stain changes hands with the cash. Denying this cultural taint by adopting the sanitized language of "compensation" is a dangerous partial truth. "Compensation" suggests that our practices of valuing a human life are purely rational, as if the payment is equal to the family's loss. "Settlement" is better, because it denotes both a capitulation and a working through: something that has been out of joint is restored to stability. "Blood money" is the best term because it carries all the stigma that comes with defusing the victim's special status in the community: when we accept the money, we relinquish our status as exiled by injustice.

Translating loss into dollar terms is the foundation of tort law, and, in the next chapter, on blood money, I look at how pricing the priceless became a hallmark of modern law. Money is symbolic: "people earmark different currencies for their many, complex, and often delicate social relations" (Zelizer 1994: xi). Blood money is especially symbolic. Because my brother had no dependents and my parents were relatively financially secure, the settlement money we received was awkward and special for my family. We did not yet see the culpability of the governments and corporations so this money seemed like a strange windfall, almost an awkward memorial to Jim. I and each of my remaining brothers received $2000 from my parents' settlement and a commitment to ongoing support for postsecondary studies. I needed money for university, but that $2,000 was almost sacred, set apart from utilitarian finances. Widows with children and parents who relied on their sons for support did not have the luxury of romanticizing the money as a memorial. Nonetheless, making the right decision worked symbolically as a parental challenge. Wives remembered their husbands best by caring for their children, and they cared for their children best by making the most secure decision they could about their economic future.

Sometimes they are even conceived historically, as if each kind of conflict resolution is in the process of replacing an earlier form: that is, as if tort replaces charity, private insurance replaces tort, and no-fault workers' compensation replaces the need for both tort and private insurance. The *Ocean Ranger* aftermath shows that this is not the case; instead, they coexist and they support one another as ways of reintegrating family members into the community.

Within days of the rig's loss, the families foundation for the *Alexander Kielland* offered support to the *Ocean Ranger* families. The *Alexander Kielland* was an oil rig that sank off Norway killing 123 men on March 27, 1980. That rig sank in bad weather and the loss highlighted the dangers of drilling in northern climates. The incredible challenges of evacuating workers from oil rigs in winter had been the focus of the inquiry into the *Alexander Kielland* disaster. Members of the St John's media put the *Alexander Kielland* representative in touch with Lorraine Michael, then Sister Lorraine, a nun schooled in liberation theology and one of the co-chairs of the ecumenical Social Action Committee, which was committed to social justice and overcoming poverty. Less than a month after the disaster, the Ocean Ranger Families Foundation started to take form as a support group composed of family members, churches and labour representatives, though it was formally structured in April (*Evening Telegram* 1982l). As the weeks wore on, the Church Liaison Committee noted "many serious problems among the families. These included at least two attempted suicides and several cases requiring psychiatric and in-and out-patient treatment" (Hawco 1983: 5). Thinking back a year later, Hawco reflected that psychological and support services were not very effectively connected with the families. Women who were not formally married to their lost man were not always treated as "kin" and were ignored by officials. Hawco notes that at least two of these women, who had children with men lost on the rig, were subjected not only to complications around eligibility for support but even questions of the paternity of their children. "In some cases… it was not just a matter of being offered support, but rather of harassment and condemnation by the victim's family. In some instances there had really been no positive rapport or relationship between the victim and his family at the time of the tragedy. However, when being hired the individual had indicated his next of kin as father or mother, for example, and this was the only legally and officially recognised griever" (Hawco 1983:6).

All the *Ocean Ranger* families received private insurance payouts of $20,000, except for the lowest paid men on the rig, the stewards, whose contract with the catering firm Atlantic Fortier Offshore Food Services did not include a policy (*Prince Rupert News* 1983). Women and children who were not reabsorbed temporarily into their parents' or in-laws' households were destitute when the *Ocean Ranger* pay cheque stopped, as it did for some, at the hour of the men's deaths (House 1987; Heffernan 2009). Some were aided through charity, either through the churches or through a fund set up by the *Evening Telegram*, the St John's daily newspaper. Money was donated from across the country as Newfoundlanders working in Alberta and Toronto collected funds. The newspaper set up a committee to administer these funds, chaired by a prominent St John's businessman. The fund received $92,558.86, only $24,000 of which was distributed to families (House 1987: 84). When the Ocean Ranger Families Foundation petitioned them for money for a handful of widows whose husbands had no insurance and who were applying for welfare, the committee administering the fund refused. The brother of one man who died explained that his brother left a young widow with three children, and all she and other Fortier widows got before the company left town was "18 roses and several letters saying the company would do what it could" (*Telegraph Journal* 1982). The newspaper's fund released seven cheques to families in need, ranging from $500 to $1,500. Lorraine Michael and the Ocean Ranger Families Foundation lashed out at the Telegram's Disaster Fund, suggesting that people should think twice before donating to it. This upset and angered the Telegram and its publisher, Stephen Herder, who felt that "the motives of well-intentioned volunteers had been questioned publicly" (*Telegraph Journal* 1982). Publisher Herder explained that the decision on distributing funds was made by a professional social worker and that the idea of the fund had been to raise money, but they had not yet determined how it should be spent. An editorial in the *Evening Telegram* raged: "What colossal nerve! What irresponsible words! The Foundation declared, rather childishly, that the Evening Telegram started its fund without asking any of those concerned whether or not this is what was wanted. Since when has any organization with charity of purpose, and out of the goodness of its heart, had to invoke such considerations?" (*Evening Telegram* 1982k). Lorraine Michael recalls:

> I have been angry and frustrated in my life but I can honestly say that nothing has ever made me as angry as that. They collected close to $100,000, and they gave out only $24,000 or something. The Foundation begged them to give the rest out to the families that needed

> it. We had eleven women—I think all of them had children—and their husbands worked for the company that did the catering. They had no private insurance and so their money stopped the minute the rig went down. They had nothing. In those days we didn't have food banks the way we do now. That has come with the development of the oil industry. Before, people went to churches and family. I couldn't get over what those women were going through. We begged and we pleaded, and they said "no." I will never understand that. (Michael 2010)

The *Ocean Ranger* Disaster Fund is now a scholarship that is open to all Newfoundland applicants.[1]

Leo Barry, a sitting member of the House of Assembly and a practising injuries lawyer, sent a letter to some families two weeks after the loss. At the time of the *Ocean Ranger* disaster, Barry was perfectly positioned to shepherd the Newfoundland wrongful death claims: he held degrees from Memorial, Dalhousie Law and Yale Law. Barry was also one of the architects of Newfoundland's burgeoning regulatory regime and a key figure in the jurisdictional disputes with Ottawa, having served as the minister of mines and energy in the cabinets of both Premiers Moores and Peckford. Described by the *Financial Post* as being "as passionate as Premier Brian Peckford" on the issue of Newfoundland jurisdiction over the offshore, Barry was part of a generation of ambitious men determined to maximize the benefits of the offshore for Newfoundlanders (Anderson 1980). Barry fell out with Peckford when the premier shifted responsibility for the negotiations from him as energy minister, and he quit Cabinet six months before the *Ocean Ranger* disaster (*Montreal Gazette* 1981; Cadigan 2009: 266). Few would have been as able to appreciate the importance of taking care of those families, both as an immediate social responsibility and in the broader sense of keeping the Newfoundland people focused on pursuing the promise of oil. Barry's letter to families, on House of Assembly letterhead with his law firm's phone number typed into the address, expressed his and his family's sympathy: "I intend to do everything possible as a Member of the House of Assembly to see that every effort is made to discover the reason for the rig's collapse. It will also be my objective to see that the next-of-kin of those lost are properly compensated, although we all understand that money cannot begin to make up for the terrible loss which you have experienced." On the same day, "Leo Barry of Halley Hunt" was interviewed on "Here and Now" explaining how complicated the cases would be due to questions of Canadian and American jurisdictions and potential conflicts with Workers' Compensation.

Douglas Moores, another lawyer representing *Ocean Ranger* families, presents Barry's role as a key component in Newfoundland's successful management

HOUSE OF ASSEMBLY
NEWFOUNDLAND AND LABRADOR

Government Members Office
Confederation Building
St. John's, Nfld.
A1C 5T7
Telephone (709) 726-6124

March 1, 1982

Mrs. Norrene O'Neill
Fermeuse
Southern Shore, Nfld.
A0A 2G0

Dear Mrs. O'Neill:

Like every Newfoundlander, I was shocked and deeply saddened to learn of your recent tragic loss in the Ocean Ranger disaster.

Please accept the deepest sympathy of myself and my family.

I intend to do everything possible as a Member of the House of Assembly to see that every effort is made to discover the reason for the rig collapse.

It will also be my objective to see that the next-of-kin of those lost are properly compensated, although we all understand that money cannot begin to make up for the terrible loss which you have experienced.

Please feel free to contact me if there is anything at all which I may do to be of assistance.

Sincerely yours,

Leo Barry

Leo Barry, Q.C.
M.H.A., Mount Scio

Letter to Noreen O'Neill from Leo Barry, MHA *(Letter courtesy of Noreen O'Neill)*

of the disaster's social effects. Moores emphasizes that the lawyers clearly saw their responsibility as "social as well as legal."

> As soon as the *Ocean Ranger* sank ... almost immediately, the next day or two, families were bombarded ... We in Newfoundland in the legal community had never been involved in what the Americans would call a mass disaster. As soon as this happened, lawyers started to de-

> scend on St John's from all over the United States, from Texas from New Orleans from New York, lawyers who had been involved in mass disasters—oil rig disasters here, plane crashes there—things of that nature, came to Newfoundland trying to represent the widows and the families. That's just the culture in the United States when it comes to that kind of work. We had never been exposed to that before. And of course that at the time that this happened ... and bodies were still ... the search was still going on ... men were being taken from the sea ... I mean it was just unbelievably traumatic here. There was no thought of closure. I mean people were still hoping to find people who had been lost at sea. That took three or four days .... Widows were getting knocks on their doors, "I'm Joe Blow I'm here from Texas to represent you. Your husband was lost. Your son was lost. Your boyfriend was lost. I'm prepared to represent you. Give me forty percent of what you get and I'll take care of you." So there was a lot of this going on. So people were inundated if you will. The legal profession didn't look very good. And this wasn't the Newfoundland lawyers, it was just the way the Americans do business in these types of cases and that's just the way they do it. (Moores 2011)

Families were "hounded" by American lawyers and the Ocean Ranger Families Foundation spoke on their behalf, advising that they "were looking at how the families should proceed with legal action" ("Here and Now" April 4, 1982). One of those lawyers telephoned my parents and they turned him down, opting to move quickly and independently with a local Nova Scotia lawyer. American lawyers approached Craig Tilley's mother, Pat Ryan, as she worked in her beauty salon in the Newfoundland Hotel: "It was incredible. They swarmed" (Ryan 2010). The Families Foundation announced publicly that the families would deal with local lawyers as a group. Leo Barry was central to the Ocean Ranger Claimants Committee, as he represented fifteen of the families and served as a lead counsel, working closely with the American lawyers who the Newfoundland lawyers collectively engaged for the case. Barry reportedly received $750,000 in fees once his clients' suits were settled. Despite later politicized attempts to cast aspersions on Barry's letter to families of March 1, 1982, he was cleared by the Bar Association of any suggestion of unethical soliciting (*Montreal Gazette* 1985).

Fees for injuries and wrongful death suits are a tricky business, and every family I spoke to was shocked by the final amounts paid out to lawyers. According to Moores, however, this could have been even more shocking: the American lawyers initially wanted a 40 percent fee, but the Newfoundland lawyers negotiated them back to a portion of an overall fee to families of 30 percent.

Despite some public misunderstanding and misinformation, contingency fees are a longstanding tradition in tort law. Negotiations are expensive, and bereft families rarely have the money to cover lawyers' hourly fees. Typically, families engage a legal representative who agrees to assume the costs of the negotiations on speculation that the settlement will both cover their costs and pay them for their work. As a Houston lawyer who has represented both large companies and injured workers and their families puts it, for tort to work, the lawyers must have "skin in the game," that is, they need to have a direct, monetary interest in the outcome (Spagnoletti 2010). In the *Ocean Ranger* cases through the Claimants Committee, American lawyers received 16 percent of the final settlement while Newfoundland lawyers received 14 percent.

The legal possibilities were dizzying and the probabilities were unfathomable.[2] Two questions that were considered in more socio-political terms by the Royal Commission also had important ramifications for the lawsuits: Was the *Ocean Ranger* a workplace or a ship? And who had jurisdiction over the offshore? If the *Ocean Ranger* was a workplace, then the *Workers' Compensation Act* should apply. That would mean that families could receive compensation but would be unable to sue. If the *Ocean Ranger* was a ship, then federal legislation applied. If Canada and not Newfoundland was the "coastal state," then the deaths fell under the *Merchant Seamen Compensation Act*, which provided compensation if (and only if) no other compensation applied. One legal scholar, writing before the Newfoundland amendment to its *Workers' Compensation Act*, laid out the mind-boggling tangle of possibilities—reaching back to judgments made by Viscount Haldane at the turn of the century—and determined that there was no simple answer to the question of which compensation system applied to the *Ocean Ranger* dead (Hayashi 1983: 176).

Two more questions complicated the legal debates around the *Ocean Ranger* cases: Would *Ocean Ranger* families win access to federal court in America? If they did not, would recourse to Texas state court be worth the risk? If families could sue in American federal court, they might still be limited to the recovery of "economic loss," but the calculus for this was much more generous in the U.S. and there was the chance of awards intended to punish the companies. If they had to go to Texas state court instead, they would go before a jury, with utterly unpredictable results.

SETTLING BEFORE THE YEAR WAS OUT – "AMICABLY, WITHOUT ADMITTING LIABILITY"

Ten months after Jim's death and before reports from any of the inquiries into the causes of the disaster, the combination of mammoth corporations—Mobil Oil, ODECO, Mitsubishi and Schlumberger—paid my family $25,000 to "settle

Jim Dodd, c 1980 (Courtesy of Susan Dodd).

and adjust their differences amicably, without admitting liability," as the contract put it. "It was like haggling over a car," my mother said. Not only did this capture the sleazy feeling of negotiating a dollar figure for a life, it also emphasized the sense that the amount offered was calculated within a vast range of contingency, according to hidden values and strategic posturing. However much the textbooks might like wrongful death settlements to be based on precedents that provide a calculus determining lost income minus monies the man would have taken from the family budget had he survived, negotiating the *Ocean Ranger* settlements was much more complex and power-laden.

Two key legal principles drove this monetary evaluation of Jim's life. First, it was held that financial settlements for wrongful death should be calculated by estimating how much the dead would have contributed to their surviving families. A single man with no dependents, the son of relatively financially secure parents, Jim was not worth much—in dollars and on textbook charts—to our family. I always joke that given my never-ending stay in university, if I had been killed, my family should have had to pay those responsible for taking me off the family's balance sheet.[3] A second key principle was that—unlike U.S. courts—Canadian courts have not traditionally ascribed financial values to the pain and suffering of a dying person or the loss of society, care, guidance and companionship suffered by the rest of us as a result of the loss. Such awards involve calculating "hedonic goods" in financial terms. They are highly subjective and are awarded most often in U.S. state courts with full jury trials. Jury trials—especially in Texas—are notoriously unpredictable, broadening the range of possibilities from "windfalls" to "inexplicable denials of remedies"

(Weinstein 2009: 1267). Canadian judges have followed the British tradition, seeing this as an almost barbaric gambling over loss. How can we estimate — in dollars and cents — a just compensation for the terror those men felt? Who was more terrified: the men who lasted an hour in a lifeboat or the men who jumped from the rig into the raging, debris-filled water? Who lost more, the daughter of a happy couple left with her bereft mother or the daughter of a miserable couple left alone with an alcoholic mother? We might be able to calculate the lost persons' probable contribution to the household in practical ways like childcare, yard work, car maintenance and the like, but how can we really quantify the loss of a parent, child or spouse?

Calculating what our family lost in Jim's earnings was relatively clean, and yet, even in my brother's case, my parents were awarded "double previous Canadian awards," as the lawyer explained to my parents. What was this? Some kind of strange gift? My parents could not know that ODECO would be shown to be grossly negligent, nor that Canadian criminal and regulatory law were not capable of punishing such offenders. They were approached by an unknown tort lawyer who encouraged them to join him and a few other families in pursuing suits against Mobil and ODECO in their home jurisdiction of the U.S. My parents attribute their refusal to sue to disgust with the whole process of assigning a dollar value to Jim's life, particularly because he was not providing for anyone. They still had me at home, a relatively difficult teenager trying to finish high school, and they felt they needed to settle quickly so we could all move on. "Well, and to be perfectly honest, your mother just couldn't take any more," my father explained. Families' psychological limits add to their often vulnerable economic situations to make them much more anxious to settle than the companies are. My parents also recalled fondly the young shore manager from Schlumberger who telephoned them through that night, through the long hours as the "rescue" efforts faded to "recovery," and "recovery" became "lost at sea." Two representatives from Schlumberger came to our house in Berwick. "They were well put together and very respectful," our family friend put it "and they just laid out what the company was doing and why." Basic civility, family pressures and the fact that my parents were far from the collective effort in Newfoundland all contributed to our early settlement.

No widows from Newfoundland settled independently, without the benefit of group support and American legal counsel, but I do not know if any of the family members from other parts of Canada did. If it happened to my parents, who did not need the money and so would have pursued a larger settlement only for punishment or reform, it could have happened to a young wife with children who desperately needed the money. Companies invest in out-of-court settlements in part for this very reason: out-of-court means off the record, written out of history. The details of such settlements and how they were determined simply disappear with time. When I first asked my parents about our financial settlement, they could not piece together a recollection of the negotiations. *Ocean Ranger* families negotiated financial settlements for wrongful death in fraught contexts. Media frenzy, emotional pain and family readjustment all complicated the negotiations for "Judas money" as one lawyer called it. Represented by a Nova Scotia lawyer, who probably spent most of his time doing wills and house sales, and not as part of the group in St John's, my parents settled before any of the inquiry testimony broadcast evidence of the shoddy management of the rig. My parents' lawyer did not return my mother's call about this, and my parents cannot remember how it all happened. Mom recalls knowing that if you want to make as much money as you can, you should never just accept the company's first offer; but they did not care. They just wanted it to end.

OCEAN RANGER CLAIMANTS COMMITTEE – "ON THE BACK OF HER DEAD MAN"

Noreen and Paschal O'Neill lived in Fermeuse, about an hour from St John's, with their nine-month-old son, Ashley. After Paschal's death, Noreen made the wintery drive into St John's for the first meeting of the Ocean Ranger Families Foundation. She recalls: "I just needed to be around people. I needed company from other people. I don't know why I went, really." She does not remember much about those early meetings, about the negotiations that led to her final settlement, or even much about the difference between her settlement and anybody else's. The Ocean Ranger Families Foundation became a forum for support and increasingly a lobby group for safety in the offshore. Only very accidentally did it also become a forum for discussing the pros and cons of offers from the companies.

Both Canadian and U.S. legal systems made adjustments in light of the *Ocean Ranger* case. In Newfoundland, the Workers' Compensation payments supported wives and children while lawyers explored the prospect of gaining access to American courts. In July, the minister of labour and manpower announced that people would be able to sue and collect Workers' Compensation benefits as they went through the negotiations towards settlement ("Here and Now" July 2, 1982). The Newfoundland and Labrador House of Assembly amended the legislation that usually blocked fault-based lawsuits following payments from the no-fault Workers' Compensation program. Though the compensation payment would be a "salvation financially" for some families (Hickey 1983), the months between the loss of the rig in February and the change to Workers' Compensation in July must have been very long indeed, especially for the families of the men who worked for Atlantic Fortier Offshore Food Services.

Their last pay cheque covered only the eleven days they worked—and not the twenty-one-day shift that Mobil and ODECO paid in full to their employees. They were not covered by a private insurance policy so they received nothing to tide them over. In mid-July, Lorraine Michael of the Families Foundation indicated that about ten families were "experiencing severe financial hardship" (Chronicle Herald 1982).

Again, Leo Barry was a central player. According to Moores, "I'd like to pay special thanks to Leo Barry. He's had a long legal and political career. Through his efforts and abilities, Leo was able to convince the Workers' Compensation people to agree to provide benefits while we pursued these claims. Without that, it would have been a much more difficult case to pursue, without that support of the Workers' Compensation Commission" (Moores 2011). The Newfoundland lawyers' decision to band together had both social and strategic advantages. The lawyers representing the families met together and formally invited American international tort experts to be interviewed. Eight or ten firms "made the trek to St John's," and from those the Newfoundland lawyers chose a law professor from Tulane University in New Orleans and firms from New York (Mobil's home) and New Orleans, as well as Benton Musselwhite from Texas (Moores 2011). In this way, Newfoundland lawyers were able to sustain some control of the situation as well as benefit from an international legal trend in the late 1970s and early 1980s: the rise of a counter-expertise to the company's lawyers in the international ambulance chaser, otherwise known as "the new elite plaintiffs bar," that is, experienced, well-resourced lawyers specializing in representing the victims of mass disasters (Wagner 1986: 44). One legal journalist described these lawyers as "lawyers who have tremendous legal and managerial skills, and the personal financial resources to litigate cases that may pay off in ten years—or not at all. They bring a gambler's instinct to cases that push against the frontiers of science and the law" (Wagner 1986: 45). As Leo Barry explained on "Here and Now," the cases would be complex and require expensive international expertise (March 1, 1982). Part of Newfoundland's "growing up" would be meeting the challenge of these international lawsuits.

Maritime law is a labyrinth of ancient precedents, international treaties and domestic laws that further complicated the *Ocean Ranger* case. If the families sued in the U.S. under the *Jones Act*, which provided for the families of mariners lost at sea, they would be limited to "economic" damages. This was the same principle as in Canadian courts, with the twist that American awards tended to be much more generous, even when restricted to "economic" damages based on a calculation of lost earnings. Even the possibility of access to American court raised the spectre of moving beyond strict "economic" damages into awards intended to punish the companies ("punitive damages") or to compensate for social losses (loss of "care, guidance and companionship") or even for the pain and suffering of the men in their final hours.

In the U.S., courts were at the beginning of what would become a trend of ruling against workers injured in American enterprises abroad who tried to have their cases heard in their employers' home country. Claims filed by American families in America were reported in the *Evening Telegram*, and they were astronomically higher than any Canadian claim would be, ranging from \$314 million (for Michael Watkins) to \$2 million (for Gerald Vaughn) (*Evening Telegram* 1982f; *Oil and Gas Journal* 1982b). As U.S. companies, particularly in resource industries, moved further and further from home, American courts became less and less amenable to holding American citizens and companies accountable in American courts (Spagnoletti 2010). American representatives for the *Ocean Ranger* families hoped to buck the trend, pointing to cases where non-U.S. citizens were killed in an accident in the North Sea, and ongoing legal processes that were underway concerning deaths from an airplane crash in Portugal. An *Ocean Ranger*-driven amendment to the *Jones Act*, which prevented foreign compensation claims in U.S. courts for workers on U.S.-owned ocean oil rigs, was consistent with this trend. As one senator said, "the United States should not continue to export its remedies for foreign workers' work-related injury claims which arose in foreign waters" (*Globe and Mail* 1982b).

Two weeks after the *Ocean Ranger* sank, Liberal MHA Stephen Neary attended debates in Washington about an amendment to the *Jones Act* (HR4863) to block Canadians and other foreigners from filing suits in America even when wrongs were perpetrated by American managers on American ships to profit American shareholders. Addressing the Newfoundland House of Assembly about the amendment to Workers' Compensation, Neary insisted that, had he not been there to shame them out of it, Congress would have bowed to pressure from the oil lobby to make this change retroactive, specifically to block *Ocean Ranger* claimants from even threatening to gain access to U.S. courts. The change was not retroactive, but the *Ocean Ranger* cases were still blocked from American federal court by Judge Robert Collins, as we will see.

Three months after my family settled, the U.S. National Transportation Safety Board (NTSB) found that, as the chair put it, "Underlying this whole thing is a management failure." Or, as one of the five NTSB members said, "the people who lost their lives may have been set up by management policies" (*Fredericton Gleaner* 1983). The board found that "ballast control-room operators on drilling rigs 'generally have little or no maritime background in ship stability or other marine related subjects,' the report said, and the *Ocean Ranger*'s master, who was responsible for the vessel's stability, had received no training from ODECO and had been aboard the *Ocean Ranger* only three weeks" (*Globe and Mail*

1983). By now, this was common knowledge in Newfoundland, but the formal statement by the NTSB made it a legally admissible reality. When I asked Doug Moores if this had any impact on the negotiations over settlements, he said: "No. Not really. If it had gone to trial that would have been helpful. Liability was not an issue. These people never denied that they were responsible. But we were not going to get punitive damages.... We weren't going to get into potential punitive damages. You know against these fellows in Canada. That was not going to happen. Now, if we had got jurisdiction in the United States and were able to argue the issue we might have been able to get punitive damages" (Moores 2011).

By August 1983, there was talk of settling without attaining American jurisdiction, though the lawyers had started legal actions in both the Federal District Court in Louisiana and in the State Court of Texas. There had been cases where the Federal Court declined jurisdiction but the State Court upheld the non-American's right to be heard before a jury (Musselwhite 2009; O'Dea 1983).

In mid-October 1983 a group settlement was very close. Newfoundland lawyers announced that a calculus had been developed by the companies and the Claimants Committee, and the settlements were being worked out. "We've reached a formula as a basis of settlement of all the cases which lawyers for both sides are prepared to recommend to their clients," (Leo Barry, in *Prince Rupert News* 1983). Lawyers used a previously scheduled meeting of the Ocean Ranger Families Foundation to present the calculus for the families' consideration. Media coverage discomfited many family members as they grappled with the question of whether to settle or to continue fighting. Throughout this narrowing the families' voices to financial terms, the Ocean Ranger Families Foundation struggled to keep its critical agenda broad and publicly recognised as such. The executive administrator of the foundation, Cle Newhook, tried to shore the families up against increasing social pressure. Whispers grew that greedy wives and mothers were "making money off the backs of their dead men." Newhook emphasized the breadth of the families' mandate, reminding the media that the families were writing a qualitative account of their experiences and researching the overall effects of offshore work. "Mr Newhook says he believes the best way for the group to ensure its future effectiveness would be to open the Foundation up to membership from the general public" (Warren 1983: 19). It is remarkable that Newhook felt the need to state the point expressly: the families' main concern was for safety, not pay-offs.

The families' decision was difficult, given that the American lawyers clearly thought families should hold out and wait for further rulings on the question of American jurisdiction, while the Newfoundland lawyers favoured the settlement, emphasizing that the October 1983 package included a release from repayment of Workers' Compensation benefits, which was a significant benefit—$60,000 or more in some cases. The companies' lawyers threw in a 10 percent "bonus" for settling in Canada, while Newfoundland lawyers raised concerns about the uncertainty of outcomes in American courts. The settlement formula was "the best I had ever seen," according to Justice Hickman. One reporter explained: "If accepted, individual settlements will range from $40,000 for parents of unmarried men, to as much as $1 million to widows of the higher-paid of the crew members. The bulk of the settlements would be in the range of $300,000–$500,000. Of that amount, about 30% would go to lawyers' fees, split 16%–14% between the U.S. and Canadian firms respectively" (Zierler 1983a).

Everything hinged on Judge Robert Collins in New Orleans, which was ODECO's hometown. The *Financial Post* reported: "Negotiations toward a settlement of the Newfoundland cases began only after Judge Robert Collins ruled in New Orleans in June that Canadian claims have no standing in his court. But Collins is now reconsidering that ruling after hearing new arguments in August, and is expected to hand down a final decision shortly. Meantime, relatives' lawyers have launched another U.S.$1 billion action against ODECO and Mobil in a Texas state court which operates under different jurisdiction principles" (Zierler 1983a). Families who rolled the dice could find themselves in one of two scenarios: in U.S. Federal Court, 5th Circuit under Judge Collins in ODECO's hometown of New Orleans, or in Texas State Court in front of a jury. Federal court was more predictable, while Texas court offered the big gamble which could pay out a windfall or nothing. Doug Moores remembers the preparation for the argument for jurisdiction in U.S. federal court as intense: the key legal issue was whether they could show "connecting factors," which there clearly were between ODECO's head office in New Orleans and the day-to-day operation of the rig. A handful of the Newfoundland lawyers went to New Orleans, along with the American representatives, to be heard by Judge Collins.

> We were told interestingly enough in an anecdotal meeting we had before in our hotel in New Orleans by a person I don't remember who it was who told us, "Listen, I don't think you're going to be very successful here," because Judge Collins was not going to be very interested in finding against ODECO because that was a very powerful New Orleans company and it would be unlikely that we would succeed in our claim against ODECO and get jurisdiction of that case in the United States. (Moores 2011)[4]

Almost thirty years after the *Ocean Ranger* loss, Cynthia Parsons-Walker and I drove up Signal Hill in St John's following months of email correspondence. She brought her envelope of important papers. Among her legal letters was a TV magazine with a headline about the salaries earned by the chair and staff of the *Ocean Ranger* inquiry. Pamela Ewing from the TV soap *Dallas* gleamed on the magazine's cover: the promise of oil, indeed. The correspondence between lawyers from Park Avenue, New York City and Duckworth Street, St John's, was on creamy linen stationery. Cynthia smiled and said, "Yes. It is beautiful. And we paid for it." The correspondence was dizzying, to-ing and fro-ing about possible access to the courts of America—Federal Court in Louisiana, State Court in Texas or California, offers and advice from several lawyers, with the American lawyers politely contradicting the Newfoundland lawyers.[5] "Conflicting statements by U.S. and Canadian lawyers, both representing the next of kin, have not made the waters less muddy. The Canadian lawyers appear to be leaning toward settlement, but their U.S. colleagues continue to encourage the victims' families to stay in court, where they can argue for punitive damages against the companies as well as compensation for their relatives' loss of income" (Zierler 1983a).

Were Newfoundland lawyers habituated to being risk-averse, given their lack of experience with mass disasters? Were they so focused on the social responsibility of caring for the families that the punitive aspect of the suits seemed less important? Was this simply a case of each set of lawyers expressing the legal culture of their own nation? Americans use lawsuits to modify corporate behaviour, but Canadians regulate. According to the Newfoundland lawyers, the difficulties of managing legal costs and predicting the outcome of the case worked strongly in favour of settling at this point.

Cynthia Parsons will never know what would have happened if she had continued on. She showed me the booklet outlining her options. Option A offered a monthly income, providing $28,000 a year for twenty years, with a $250,000 lump sum at the end. Option B offered a monthly income, providing $23,883 a year for her life, with a $250,000 sum after twenty years. Option C offered a monthly income, providing $17,751 a year, for the rest of Cynthia's life, but increasing by three percent a year and with no lump sum. In addition, there were structured settlements for each of the two children for roughly another $10,000 a year.

Cynthia Parsons chose a "structured settlement" (Option B) that paid out over her lifetime. What this meant for Cynthia Parsons and her children was a cheque for $74,221.09 on January 5th, 1984, and an investment of $250,000 that would pay about $30,000 a year. This reflected the total settlement of $502,300, minus fees paid to American lawyers ($79,728) and to the Newfoundland lawyers ($69,762). The settlement also included a release from

Clyde Parsons and his children (Courtesy of Cynthia Parsons)

repaying the Workers' Compensation that Clyde Parsons' family received while pursuing their lawsuit against ODECO. In a sense, then, the Newfoundland taxpayers supported the Ocean Ranger families directly in order to secure the best possible settlements from the companies. One of the added pressures to settle in December came from the fact that as of January 1, income from invested settlements would no longer be tax exempt (*Calgary Herald* 1983c). Clyde Parsons' family was left with a solid income of $50,000 or so a year. This was good money in Newfoundland in 1984, but it was more like getting a civil service job than winning the lottery.

Families who took lump sum payments would have had a more uneven experience, as there were significant amounts of money involved, and for many people, these lump sums were worth more than all the money they had ever seen in their lives (Moores 2011). Court documents show that the family of one Atlantic Fortier Offshore worker received $146,000 in immediate cash, plus an invested premium of $100,000, which paid out $10,000 a year for five years, and then that $100,000 was repaid to the family as a lump sum in 1999. As Noreen O'Neill's letter explained the decision to settle in Canada instead of pursuing the companies into U.S. jurisdiction: "This proposed settlement is certain and immediate. While you may receive a larger award in the United States Courts there is still the possibility that the U.S. actions will be unsuccessful. In that event you can expect to receive a substantially smaller award in Canada that will probably be three or four years down the road" (O'Dea 1983). Chief Justice Hickman announced the settlements for wives and children from the Supreme Court (O'Neill 2010; *Toronto Star* 1983). As both a Supreme Court Judge and the chair of the Royal Commission, Hickman embodied the legal management of the potential crisis in confidence in government and economy. The December 1983 settlements ranged from $250,000 to $900,000 for men with dependents and were around $40,000 for single men with no dependents.

Moores wants people to remember that the *Ocean Ranger* disaster was in the early days of oil rig work, when many of the men were making very modest amounts of money. He is proud of the ways the Newfoundland lawyers managed the determination of the settlements:

> We came up with some, let me put it this way, some novel work that was performed.... These young men were working at sea. They were three weeks on three weeks off, or two weeks on two weeks off, whatever. And what we tried to do was to try to say that they had an income when they were working at sea but when they were home they were barbers, they were cutting hair, they were driving taxis, they were freelancing. And in lots of cases the fellows who were doing this were working for cash. So we had to try if we could as best we could, we were able to escalate the value of the economic loss by saying that while Mr Smith only would have earned $25,000 when he was at sea, at home he would have earned another $15,000 on an annual basis so his total annual income was $40,000 a year. (Moores 2011)

Of the fifty-six Newfoundland families, all but three appear to have settled through the Ocean Ranger Claimants Committee in December 1983. Risking denial of jurisdiction in American Federal Court and then a jury trial in Texas was a gamble that young widows with children could not afford. Justice Hickman recalled that the settlements were struck about half way through the inquiry. As for the pursuit of access to U.S. jurisdiction, Justice Hickman thought this unnecessary: "That was what we referred to as 'Texas law'—the interpretation of Texas law is so wild that that's the place to go if you haven't got a claim" (Hickman 2008).

The Newfoundland political establishment took care of the *Ocean Ranger* aftermath in the double sense of both making sure the oil companies paid enough to keep families off the provincial welfare rolls, and managing the potential legitimation crisis by restoring public confidence in the promise of offshore oil development. A sitting MHA and former energy minister took the lead in the group negotiations over settlements and in engaging American tort specialists. The House of Assembly amended the *Workers' Compensation Act* so that bereft and destitute families could negotiate settlements that could support them for their lifetimes. There was a commitment to community that played out in a kind of paternalism but also in a wakening sense that Newfoundland's legal community had to step up to meet their new international challenge. Nonetheless, the English tradition of approaching financial awards for loss of life with caution and even revulsion carried over into the legal thinking about the *Ocean Ranger* settlements. This is exemplified for me in an anecdote Chief Justice Hickman told me about the first case he ever won.

In 1950, shortly after he was called to the bar, when he twenty-two years old, Hickman represented the widow of Jimmy Foley, a man who had been run over by a drunk driver. The police had picked that driver up before the accident, noted that he was intoxicated and then let him go. That driver went on to kill Foley. The jury in the manslaughter trial "let the guy go on the basis of sufficient doubt," Hickman recalled. In the civil case, however, Foley's widow won. Her award was $5,000. In a less sensational version of the O.J. Simpson outcome, the courts recognized that they could not draw sufficiently clear lines around the driver's intentions and actions to justify a criminal charge, but they could see some level of culpability that could be expressed in a financial settlement. Later, as he was striding along a Halifax sidewalk feeling proud of himself, Hickman heard a senior Halifax lawyer calling out to him from the other side of the street: "Hickman! Nobody's worth $5,000!" (Hickman 2008).

ROLLING THE DICE – "WHO WERE THEY TO PUT A PRICE FIGURE ON OUR SONS' LIVES?"

The tensions between Newfoundland's rich culture and its struggling economy, complicated by the aggressive individualism of oilmen from Alberta, Louisiana and Texas, carried over into the legal negotiations that assigned dollar figures to the lost men. At least three families from Newfoundland continued to pursue access to U.S. jurisdiction. ODECO sued the two most outspoken—Patricia Hickey and Patricia Ryan, each of whom lost an adult son in the disaster—for breach of contract. The Bursey family pressed on to the point when it seemed likely that Mrs Bursey would have to testify before a jury in Texas. Outside of Newfoundland, some families, in particular, Scotty Morrison's family, pursued their claims well into1984.

According to his brother Brian, Paul Bursey's mother settled only when it was clear she would have to travel from St John's to Texas and appear in court

Paul Bursey, electrician, at work on the Ocean Ranger, *1981 (Courtesy of Brian Bursey)*

before a jury to testify about her son's central role in her life. Brian had taken a full summer to put together the file demonstrating his brother Paul's central role in their family, both as a companion and as the key support for his ageing mother. That file was intended for the judge and jury of the Texas court. Brian included original photos, letters and sentimental notes that disappeared into the lawyers' offices forever.

Due to Scotty Morrison's career in the National Hockey League, the Morrisons had more experience with media frenzy and more legal resources than most families. Morrison remembers: "A good friend of mine was on the board of directors for the Toronto Maple Leafs and he was a lawyer himself... and he said, 'Well, Scotty, leave this with me,' and he put me in touch with this Lou Paisley [an experienced lawyer with no other connection to the *Ocean Ranger* case]. They thought that since this was going to be an American case, it would be much better to be represented by an American lawyer" (2010)

On November 16, 1982, the Morrisons received an offer of $20,000 to settle for Perry's death. Their U.S. lawyers responded with strategic vigour, as the letter from the New York tort specialist to Morrison's lawyer states:

> In reference to the letter from Mobil's attorneys, I would like to give you my reaction. Many of the recent disasters have brought forth let-

> ters of this sort seeking to sound "reasonable" about settlement offers and attempting to reach a grieving family before a lawyer is retained on a contingent fee.
>
> It is true that a big fight is shaping up on retaining jurisdiction in the United States on the Canadian death cases. However, we have a very good chance to keep the cases in the United States, since we have a U.S. flag rig, the owner doing its principal business in New Orleans, and many Americans are suing, as well as Canadians. The rig was inspected by the U.S. Coast Guard and subject to American law. On this basis I believe the offer of $20,000 Canadian is very low and I would recommend to the family that it be rejected. (Kreindler and Kreindler 1982)

The complaint filed in January 1983 in New Orleans against ODECO on the Morrison family's behalf claimed "sustained damages in the nature of loss of support, contributions, subsistence care, counsel, attention, consortium, society, inheritance, nurture, guidance, mental and moral training, services and all other such benefits, all of which damages amount to $3,000,000" (*Morrison v. Ocean Drilling and Exploration Company et al.* 1983). The complaint continues to claim that Perry's death and the damages done to his family were caused by ODECO's negligence. Furthermore, it argues, because ODECO was "guilty of wanton and wilful misconduct" the family should recover punitive damages of another $3 million. Scotty Morrison's legal correspondence clearly shows that the finding of the National Transportation Safety Board of ODECO's management failures were useful to the families' lawyers.

That said, by the following November the Morrisons were abruptly advised to accept $40,000. Both Scotty and his own lawyer, Lou Paisley, were startled by this advice. The response from the tort specialist in New York was unequivocal: the American judge had rejected their claim be heard in the Louisiana court. Appeal was possible "but it is always a problem ... asking a judge to change his mind" (Kreindler and Kreindler 1983).

Agreeing to the settlement was a painful decision, all the more so because it was, in the end, a practical one: in Canadian court a single man with no dependents is literally worth nothing to his parents. The presupposition is that he would not have contributed financially to his parents' household. No other losses are quantifiable, according to the logic of "replacement of lost earnings." The Morrison family did not care about maximizing their financial gain. They were striving for an expression of justice—a kind of justice that Canadian civil lawsuits do not and have not been intended to deliver. Scotty Morrison saw that the companies should be punished, and he clearly realized that, as things, stand, there is no way to do this in Canada. The closest approximation of punishment was in trying to get punitive damages through the American court system. As we will see, this too is inadequate to Canadian expectations of justice.

Though he swore he would never settle, when it became clear that American federal court was closed to them, Scotty Morrison decided to drop the case in Texas: "My attorney said to me, 'They are going to make life so miserable for you. Do it. Get it over with. Get on with your life. That's my professional advice to you,' and so that is what I did" (2010). Morrison angrily read the letter from the New York law firm Kreindler and Kreindler laying out the reductions to their gross payment of $40,000, which ends: "I figure the proceeds are roughly as follows: recovery $40,000 Canadian—$30,800 U.S., expenses [incurred by other law firms] net recovery $27,239, total recovery to you and your wife $18,967." The Morrisons' cheque was less than my parents' settlement, even though the Morrisons' award was almost double ours.

Pat Ryan recalled that the first lawyer who tried to explain tort's logic of replacing lost earnings made the mistake of using a pencil as an example. She remembers: "'Now let's say this pencil is your son.' A pencil! My son!" She burst into tears and refused to talk about money for another year. "But when it came to making the settlement, I was ready. I had my fighting talons up and I was ready to get that oil company because I always felt they killed my son." When the Claimants Committee members were briefed on their final settlements in October 1983, Pat Ryan and Pat Hickey shook hands under the table. Pat Ryan recalled: "I said to her, 'Are you going to sign off and let them get off that easy?' She said, 'No, I don't want to.' We shook hands right under the table." The "two Patricias," as I think of them, strike me as an unlikely team but, once in league together they were virtually unstoppable. A blaze of energy and determination, carefully coiffed and manicured, Pat Ryan is, as one of the other family members described her, "like a stick of gum." Pat Hickey was too unwell to interview for this book, though she was interviewed by Doug House during the Ocean Ranger Families Foundation *But Who Cares Now?* project, and I did have the pleasure of talking to her for an article in 1993. "How's your mother?" was her only question for me when that interview wound down. Lorraine Michael, who was then Sister Lorraine and chair of the Ocean Ranger Families Foundation and is now the leader of the Newfoundland and Labrador New Democratic Party, went to school with Pat Hickey. She was not at all surprised by her determination in the face of the international oil companies: "You wouldn't necessarily know it to look at her but she has this inner force, this quiet strength" (2010). That strength is evident in her clear-eyed vision of what happened to her son and then to her in their lawsuit: Hickey explains that she does not like cynicism, "but in this particular case it seems to stick out like a sore thumb." "Cynicism about what?" the interviewer asked. She said:

> About—us, the families, we can say what we like, get upset, make statements to the press but in the end the end result that I think of is the fact that these are the money people and we are only little termites or just thorns in their sides. They want to get on with the job of making big bucks and we are just pests to them and I think that when it comes to human dignity or human life they just don't give a scrap about that. That is how those people operate in my opinion. They have a lot of money to spend, they have a lot of work to do also, but when it comes down to the basics, human dignity and human life, I don't think they care. (Hickey 1984)

Some time after their handshake, Pat Ryan was served papers by the police while she was styling hair at her salon. As Ryan remembers it, ODECO was suing her for breach of contract because she walked away from a $40,000 offer to settle for the death of her nineteen-year-old son, Craig. As soon as she received the papers, Ryan called the CBC:

> I felt they were responsible for killing our children and I wanted to know who they thought they were to put a price on our sons' lives. Who were they to put a price figure on our sons' lives? Without even taking responsibility, liability for anything that happened. They just didn't take responsibility. They moved out of here. And that was it. "We'll pay all them off. Shut 'em up. They'll be happy." And we wouldn't let that happen. We wanted safety. We didn't want it to happen to other children, what happened to our children. (Ryan 2010)

ODECO's case against Ryan was dismissed, but it had an effect: "They actually dragged us through court. That actually made me more determined" (Ryan 2010).

Why would such a huge group of companies bother to sue grieving mothers for breach of contract? By the time the case came to court, the two Patricias and their families had pursued ODECO and Mobil in U.S. court for more than year longer than most families. What was more, they were in league with one of the most successful and controversial international ambulance chasers of his generation, the Texas personal injuries lawyer Benton Musselwhite.

Musselwhite was one of the U.S. lawyers engaged by Newfoundland lawyers. He does not remember how he heard about the disaster, only that he was approached to work with the Newfoundland committee. Musselwhite is known by some as "an international ambulance chasing scandal" and one of the "masters of disasters" (*Victoria Advocate* 1989). To put this in more genteel terms, Musselwhite is one of the powerhouse plaintiffs' lawyers that emerged through

Benton Musselwhite, Sr., Patricia Hickey, Patricia Ryan in New Orleans (Courtesy of Patricia Ryan)

the mass torts of the last three decades. Even as early as 1982, Musselwhite had a reputation for pushing litigation. His most famous case was arguing on behalf of American Vietnam veterans who sued the American government for the injuries they sustained from the military use of Agent Orange. Early in 1981, Musselwhite collaborated with another Texas lawyer to challenge the authority of the lead counsel in the mass tort case. "Musselwhite, who had been a top aide to Ralph Yarbrough, a liberal senator from Texas, claimed to have obtained retainers from more than 1,100 veterans; most of these were members of Citizen Soldier, which had lobbied for amnesty for Vietnam-era draft evaders" (Schuck 2004: 74). When the Agent Orange settlement was set at $180 million, in 1985, Musselwhite quit as co-lead counsel, as the claimants multiplied from about 20,000 to 250,000, thus watering the awards down to the point of being almost meaningless. Instead of accepting such a settlement, Musselwhite argued: "I believe we should go ahead and try Agent Orange" (Wagner 1986: 47).

From the Agent Orange case, Musselwhite travelled to Newfoundland's *Ocean Ranger* and then to Bhopal's gas leak, to Aberdeen's *Piper Alpha,* and to Panama with Dole banana workers, just to name a few of his high-profile cases. Suffice it to say that Musselwhite is no stranger to controversy. In the mid 1980s, while the victims of Union Carbide were seeking access to American jurisdiction, Musselwhite was disciplined by the Texas bar for "misinformation and soliciting." He was suspended from practice for ninety days and placed on probation. Not only was this a problem for him, it disrupted the strategic legal approach of litigants from the Bhopal gas leak case, who were trying to pursue Union Carbide into American courts. As one of the activists from that case explained:

> But we had a decision to make and we had to make it fast. And we had to make it knowing that Benton Musselwhite, the lead lawyer offering to reinitiate the proceedings in the United States, had had his professional license suspended by the Texas Bar based on allegations of misrepresentation and solicitation. Musselwhite argued that he was the object of a smear campaign by the Texas Bar, which he described as a corporatist enclave made uncomfortable by his anti-establishment challenges. For the appeal of his suspension, Musselwhite received supporting amicus briefs from the Veterans' Association which he had worked with on the Agent Orange litigation and from numerous grassroots groups he had worked with on environmental problems. (Fortun 2001: 28–29)

A court ruling from 1994 blocked Musselwhite's attempt to have his breach of probation overturned because he had flown to Scotland in the wake of the *Piper Alpha* disaster and held one of his press conferences, which some claimed directly solicited grieving clients in an unethical way. As the Court of Appeals described the practice:

> Benton Musselwhite is a Houston attorney who made a name for himself representing plaintiffs in complex personal injury cases, usually involving mass disasters. In 1987, he attracted the attention of the State Bar of Texas as the result of his public statements soliciting clients. Typically, it seems, Musselwhite held "press conferences" during which he would announce his plan of action in response to a particular tort and detail his qualifications. Not incidentally, he rarely failed to include information on how potential clients could reach him if interested in his services. (*Musselwhite v. The State Bar Of Texas* 1994)

In 1996, Musselwhite was charged with hiring a runner to solicit clients in the wake of a ValueJet DC-9 crash in Florida that killed 110 people (*ABA Journal* 1986: 32). Soliciting clients in person, or on the telephone—as my mother recalls being solicited—is considered to be unethical, if not illegal.

Musselwhite now divides his time between litigation and advocacy through

Craig Tilley (Courtesy of Patricia Ryan)

"One World Now," a non-profit that lobbies for global environmental and humanitarian law. He is expansive. When I joked that I was recording our interview so he would not sue me for inaccuracy, Musselwhite drawled: "Awwwww, I've got bigger fish to fry than you" (2010). Even addressing the United Nations he introduced himself by saying: "I make my living suing corporations all over the world" (Musselwhite no date). His accent, confidence, charm, enthusiasm for his children and his focus on social justice are balanced by aggressive pragmatism. "So, your fee is 30 percent of any settlement?" I asked. "That's right. And 40 percent if I have to put a pencil to it," that is, if it goes to court. Musselwhite could not recall the details of the *Ocean Ranger* case, though he did remember the two Patricias. He reacted strongly to my family's settlement of $25,000: "That is a dishonour to your brother's memory" (Musselwhite 2010).

In rulings from 1985, the Newfoundland and Labrador Supreme Court Trial Division found that both Patricias had instructed their lawyers to accept a $75,000 U.S. settlement on December 6, 1984. In Pat Hickey's case, she is thought to have done this almost accidentally, without consulting her husband and without realizing that she was entering into a legal contract: she expected to have to sign something and not just to say "yes" to her lawyer. Pat Ryan (then Anderson) is thought to have agreed to the settlement (which other members of her family endorsed) and then changed her mind. The judge does not press this question, however, because ODECO had undone the contract themselves by adding conditions to the agreement when they made the payment. ODECO broke its own contract, and so the suits against both Patricia Ryan and Patricia Hickey were thrown out (ODECO vs Tilley 1985; ODECO vs Hickey's Estate 1985).

The ruling in Pat Hickey's case is particularly helpful in giving an overview of how so-called negotiations unfolded. It is instructive to consider the offers the Hickeys rejected in contrast with those accepted by others.

Around the time my parents settled, nine months after the loss, ODECO offered the Hickeys $20,000. The February 1983 National Transportation Safety Board report publicly pointed the finger at ODECO and the prospect of winning American jurisdiction was very much alive. American counsel were reluctant to have the families settle at this point. Persuaded by Newfoundland counsel that the offers were "certain and immediate," as well as far higher than Canadian precedents, fifty-three of the fifty-six Newfoundland families accepted settlements that totalled $20 million in December of 1983 (O'Dea 1983f).

Craig Tilley's, Greg Hickey's and Paul Bursey's families were the only Newfoundland families holding out (as far as I know). Around the time that Scotty Morrison's family settled, by May 1984, ODECO's offer to the Hickeys doubled. In August it hit $50,000, with October bringing a "firm offer from ODECO" of $60,000, followed by U.S. $75,000 in December. On December 20, 1984, ODECO sent the Hickeys a cheque for $94,838.40 on the grounds of Pat Hickey's alleged acceptance of the U.S. $75,000.

Pat Hickey and Pat Ryan continued their suits with Benton Musselwhite. They travelled to New Orleans to hear rulings on jurisdiction. "And there we were, two little girls from Newfoundland, sitting in that courthouse in New Orleans. The lawyers would come in just to look at us" (Ryan 2010). Pat Ryan reflected on a New Orleans judge's pep talk about how good it would feel once a settlement was signed and delivered:

> She offered to take us out to dinner, and then to have a beer right there in the office while we talked about the suit. She told us about how her husband just signed off on a settlement for an injury... how liberating it was. "You will feel so good when it is over." And I said to her: "your husband is still alive, I take it? We're here because these people killed our children and nobody's doing anything about it." (2010)

Ryan insists that this was an informal chat with a judge, in her chambers in New Orleans. Ryan and Hickey were so spooked by being in ODECO's hometown that they changed their tickets and flew home earlier than they planned. "They were probably bought off by the oil companies," Ryan still says of any U.S. judge related to the case.

What are we to make of the fact that Robert F. Collins, the federal district judge who ruled that New Orleans was an inappropriate forum for Canadians to pursue suits against an American company, was convicted in 1991 for accepting bribes?[6] A drug dealer went to the FBI and told them Collins offered him the opportunity to pay for a lenient sentence. The FBI gave the drug dealer $100,000 in marked bills, "most of which was later recovered from the offices and the safe deposit boxes of Judge Collins and [one of Collins' friends]" (*New York Times* 1991). Judge Collins' lawyers argued that, because he was black, the case against him "was motivated by a desire within the Justice Department to prosecute black officials" (*New York Times* 1991). Even if Collins received unusual scrutiny due to racism, this whole story is hardly a resounding endorsement of the delivery of justice in the U.S. federal court's Fifth Circuit during Collins' tenure. Incidentally, Collins continued to receive his salary of $125,000 a year while serving his six years and ten months, until he resigned his life-appointment in 1993, when he finally faced an impeachment trial (McQuaid 1993). While none of Judge Collins' rulings on the *Ocean Ranger* case went to appeal, the subsequent events suggest that the Newfoundland lawyers' reluctance to risk the unknowns of United States federal court may have been prudent. In August 1985, when Judge Collins finally closed the door on the two Patricias' pursuit of ODECO and Mobil, Pat Hickey was quoted in *the Evening Telegram*: "'Nothing surprises me about this case, absolutely nothing,' she said, noting that the legal proceedings have been going up and down 'like a yo-yo'" (*Evening Telegram* 1985). When I asked Doug Moores if he thought any money had changed hands between ODECO and Judge Collins before their jurisdiction hearing, he replied, "I don't know. But if someone told me it had, I wouldn't be surprised. Let's put it that way. ODECO would have known Judge Collins a lot better than we knew him. That's for sure" (Moores 2011).

In the end, Benton Musselwhite advised Pat Ryan and Pat Hickey that their suit should be settled in New Orleans. Pat Ryan settled for $275,000 (to be shared with her ex-husband), the Hickey family for $250,000.

PRICING THE PRICELESS: FALSE CLOSURE – "WE WERE A NUISANCE, A NUISANCE THAT MADE NOT ONE IOTA OF DIFFERENCE"

Ocean Ranger settlements were negotiated in a labyrinth of power-plays and posturing where even a highly competent local lawyer could be overwhelmed by the international legal expertise summoned by Mobil, ODECO, Mitsubishi and Schlumberger. I wonder, sometimes, if the payouts to family members even matched the payouts to the corporations' lawyers and lobbying firms. How much were these companies willing to invest to avoid setting precedents for higher, legally imposed settlements and a juridical finding of negligence against them? The companies, too, were covered by private insurance, and their insurance companies were involved in the negotiations of the settlements in ways that were utterly hidden from public view. We will never know the exact role the insurers for Mobil, ODECO, Mitsubishi, Schlumberger and the others played, except that ODECO recovered more than $100 million in insurance payments (*Oil and Gas Journal* 1982c; *Calgary Herald* 1983c) and settled out-of-court for an unknown sum after launching suits against the rig's builders, designers and suppliers for $165 million (*Offshore* 1983: 8) as well as the lifeboat manufacturers and the insurance company for the caterers (*Oilweek* 1983). Doug Moores explains:

> That was an insurance claim all the way. At the end of the day it was the insurance companies who settled the claims, not Mobil and ODECO. I don't know what the issues were for them in terms of deductible, but the insurance companies paid the claims and we were dealing with the insurance lawyers to finally settle. Right to the bitter end, the insurance lawyers representing the big companies were there at the table. (Moores 2011)

Nonetheless, the effect of such an insurance payment, even coming on the heels of three other very expensive oil rig disasters, was expected to have little impact on industry premiums (*Offshore* 1982: 44). That is to say, private insurance supported ODECO without reforming the industry more broadly.

Sums paid to settle with families were contingent, strategic and driven by both attempts to represent the particularity of each man and also the urgency of the companies' desire to settle. In Canada, the *Ocean Ranger* settlements ranged from $25,000 to $2 million. In the U.S., the launched suits reached into the tens of millions of dollars, according to media reports, but I have no verified reports of the final settlements. Canadian claims were founded on a 1978 Supreme Court ruling's formula for determining damage for loss of life. The case *Andrews et al. versus Grand and Toy Alberta Ltd* gives a table that helps lawyers estimate life expectancy, potential income, family expenses and so on. In relatively straightforward cases, "economic loss" is calculated by estimating the amount the man would have earned had he lived a natural life and subtracting the amount he would have spent on himself. This latter calculation involves estimating the costs of his habits, the likelihood of his marriage breaking up and so on. Family members were initially alienated by the negligence of the companies and governments, then again by the failure to name and punish the wrongdoers and yet again by the process of settling on the blood price. Family members overcame their alienation, to an extent, when they accepted money. However, in order to do this they had to accept terms given by the political and economic systems. Instead of making the companies speak the language of loss, families were forced to speak of their loss in the language of finance. *Ocean Ranger* families did not sign confidentiality agreements about their settlements. They did not need to because the process of serving their men up for the calculations of lawyers and economists instilled so much guilt that their sense of themselves as public figures was radically diminished.

The guilt of settling, without inflicting any punishment or wresting an admission of guilt from the companies, still haunts all the family members I met. Cynthia Parsons-Walker said: "Looking back, I think I made the wrong decision. We were offered a package with different options and were told to go home and decide. We had 24 or 48 hours, whatever. It wasn't enough... I know we did better than they ever did before in Canada, but I felt at the end it was too quick. It was a barrage—'this is the best you will do. You really have to take it'" (in Campbell and Dodd 1993: 18). Those who settled most quickly fear they abandoned their men, while those who hung in rush to explain that it was justice they sought, not cash. The widow of Robert Wilson, an Alberta engineer and U.K. citizen who made $95,000 a year, settled in April 1985 for $2,024,229.73 (Supreme Court of Newfoundland document). Even though $2 million seems like a great deal of money, it is worth remembering that the settlement was with ODECO (i.e., Murphy Oil), Mobil, Mitsubishi, the government of the United States, the American Bureau of Shipping and, in Wilson's case, his direct employer, Neyrfor Turbodrilling. For the companies, the difference between paying out $20,000 or $200,000 or even $2 million per man was almost negligible. Settlements in the United States for Americans are thought to have been somewhere between double and ten times Canadian settlements (George Baker, in Nishman 1991: 65).

The negotiations over money diverted family members from other fights, such as the struggle to regulate industry and improve safety. When the settlement was finally reached, it did something to people. It broke them or calmed them; it brought closure that let people once again separate their family lives from their political lives, but it left its mark for good. The money that changed hands to "settle" after the *Ocean Ranger* disaster was far more symbolic than its role as a marker for economic value suggests. Money played a crucial role in restoring public confidence in the promise of oil and in turning the *Ocean Ranger* disaster into "history." In accepting that money, *Ocean Ranger* families gave up the right to force our accused to explain themselves to us. This was crucial to easing the family members out of a public role as critics of government and the companies.

"Closure" in its financial form felt like failure: it was at once humiliating and essential to re-entering life. It was a pragmatic decision for the living, for surviving children and for those in the family who simply could not take any more discussion of loss in the financial terms of charts, tables, formulas... and pencils. The meaning of the loss to a family always exceeds financial valuation. However materially essential the money is, closing the settlement entails acceptance of a relation of equivalence which some family members experience as a betrayal of the absolute uniqueness of the lost. Blood money between agents as unequal as these massive international companies and individual workers' families is not about reciprocity. Such a payment is not about a meaningful cost to a "perpetrator." Accepting money reintegrates family members into a legal exchange system and thus publicly defuses the authority of the victims, especially in their calls for blame and retribution.

Family members' social roles were depoliticized by accepting the settlements because there is a taint that comes with this kind of money. Neighbours are fascinated by it, but they take it as a final statement of sorts, and after the money has changed hands, they are less likely to see the bereaved family as having a rightful place in criticizing government and corporations. Tort and out-of-court settlements work through the alienation of family members at the cost of relinquishing their special survivor/victim status. To be a victim was a deeply ambivalent role, but it did give family members an opportunity to access politicians and the media: for a time, at least, the public was interested in the critical things they had to say about oil companies and governments. The financial settlements transformed this victim status. Family members started to lose their peculiar authority once they received their blood money. Pat Ryan said: "People are so judgemental. They crucified me. They think I got a million dollars. They still talk. They all think I got a million dollars. I didn't listen to them" (2010). The former executive director of the Ocean Ranger Families Foundation reflected:

> I think it was the money that did it. The public simply weren't able to come to terms with the fact that the wives and the parents and the children were kind of paid off by the companies. One example comes to mind. I was lined up in the bank, and one of the *Ocean Ranger* widows, as they were collectively known, was in front of me. Then, from somewhere behind me I heard, "Look at that one. She got all her money on the back of her dead husband." (Newhook, in Heffernan 2009: 182)

Blood money heals and it taints; it may heal because it taints. Negotiating and then accepting blood money for the loss of our men was deeply ambivalent for family members. Families were shocked, first by the loss of their men, and then again in a series of aftershocks, including the realization that the governments and oil companies had abandoned the men, the stigmatization of cash exchange that closed the lawsuits, and the release of a damning inquiry report unaccompanied by any punishment.

Pat Hickey put it best when I asked if she believed their actions had affected ODECO: "Not one bit. I don't think it left its mark on them in any way, shape or form. We were a nuisance, a nuisance that made not one iota of difference. I think there's allot of things still unresolved. When the press calls me on the anniversary to ask how I feel, I say, 'Rotten.' I feel rotten ten years after, I'll feel rotten twenty and thirty and forty years after" (Hickey, in Campbell and Dodd 1993: 18).

TO PUNISH, DETER AND VINDICATE RIGHTS? – "AT THAT AMOUNT, YOU WOULD HAVE TO SAY THEY GOT AWAY WITH IT"

Of the eighty-four men, fifty-six were from Newfoundland, twelve were from other Canadian provinces, fifteen were American citizens and one was British. When families negotiated with the oil companies over the settlements, they entered debates about how tort's four functions—to punish, to deter, to compensate and to recognize rights—differed between cases and cultures. The threat of a successful, or embarrassing, tort action against a large company almost always leads to an out-of-court settlement (Issacharoff and Klonoff 2009: 1200). Socio-legal literature is perplexed by tort law, generally, and even more by the relative merits of taking a tort action before a judge and negotiating a settlement outside a formal trial (Fiss 1984). In Canada, tort traditionally served less to reform systemic problems than it did in America, but the *Ocean Ranger* cases spanned both legal systems. The rig was operating in Canadian waters with a mostly Canadian crew, but it was managed and owned by a Canadian branch of an American company (ODECO) under contract by another Canadian branch of an American company (Mobil Oil Canada). Though a Canadian company legally managed the *Ocean Ranger*, the senior management were American citizens living in the U.S., and executives in head offices in New Orleans and New York made even day-to-day decisions. Consequently, the *Ocean Ranger* cases raised questions about what lawsuits were for, what the payments meant to the victims and perpetrators, and whether American courts had any business hearing cases about their citizens' behaviours in other countries.

Pricing wrongful death may be the most vexing issue in tort law. It is impossible to calculate the loss of life and love in financial terms that satisfy anybody. The rulings on payments follow different logics in different cases, even within the same jurisdictions. The awards can have radically different meanings for parties: $25,000 means one thing to Joyce and Ed Dodd on Main Street and quite another to Mobil Oil, one of the most profitable corporations in history. The medium of money does not translate into the same meaning for the parties involved. The social impact of negotiating and accepting a settlement far exceeds the contract between the parties. Under common law, a persons' claim to compensation for injury dies with him, so, absurdly, a person without dependents is worthless and an injured person is worth more than a dead one. It is like the old joke: it costs less to kill than to injure, as long as it is clearly accidental. According to tort law, people live on in their spouses and children. Their absence is quantifiable as lost income and, more problematically, in the man's lost contribution to the household in housework, home upkeep and also in family life more organically, as a caregiver and a gender role model.

Media frenzy, emotional mayhem and family readjustment all complicated the already complex negotiation of blood money. Lawyer Ray Wagner represented the Westray Families Group as they pursued the possibility of suing the Nova Scotia government for its failure to regulate the mining company that ran the fatal Westray coal mine. Wagner insists that the cultural discomfort around blood money is exacerbated and sustained by deliberate and constant public relations campaigns that denigrate anybody who receives financial compensation for injury or loss. In the wake of the *Ocean Ranger* loss, both symbolic and economic aspects of blood money worked to re-legitimize political and economic systems. Blood money reintegrated family members and it silenced them as public critics of the companies and government. Negotiating and accepting blood money also repeated the trauma of the initial loss.

The Newfoundland lawyers took care of families in the wake of the *Ocean Ranger* disaster. Their ability to operate as a tight community and to draw on international expertise without being overwhelmed was impressive and facilitated, of course, by the strong sense of community in and around the families. Moores was particularly emphatic that their ability to adapt to meet the new challenges of this mass disaster was related to their ability to control the story: "It was very important to us to make sure that there weren't all kinds of yarns going on out there. It would have caused a lot of trauma to people who were grieving the loss of their husbands and boyfriends and children and brothers" (Moores 2011). The economic loss was compensated and at unprecedented rates for Canadian wrongful death settlements. As for tort's other stated objectives: to punish, to deter and vindicate rights? In the *Ocean Ranger* aftermath, the massive companies paid a total of $20 million for the initial Claimants Committee settlement plus an undetermined amount to families who settled on their own. Recalling that ODECO received insurance payments and its own financial settlements from the other corporations, it is hard to see how the responsible agent was even touched financially by the disaster. Whatever those settlements to families were, the payments of between $20,000 and $2 million per man were no punishment for ODECO, the subsidiary of oil giant Murphy Oil, or its partners. Purchased in 1991 by Diamond Offshore Drilling, ODECO, the former subsidiary of Murphy Oil, is remembered as a drilling pioneer. The Wikipedia entries for both ODECO and Diamond Offshore Drilling note: "ODECO's rigs continued to rack up 'firsts' in the industry in the 1970," with no mention of the *Ocean Ranger* disaster of 1982.

St John's personal injuries lawyer Ches Crosbie is engaged in a sustained public education and lobbying program to extend Newfoundland tort to include awards for losses that exceed the strict calculation of economic loss. He is particularly adamant that we need to improve the capacity of Canadian courts to use tort law to "modify the behaviour" of negligent corporations through awards for punitive damages. Looking back, from a thirty-year vantage point, when I pressed the question of whether a $20 million settlement package paid by the insurance companies of ODECO, Mobil, Mitsubishi and the rest could in any sense be seen as contributing to their "modified behaviour," Crosbie says: "at that amount, you would have to say they got away with it" (Crosbie 2011).

NOTES

1. The original terms of the *Ocean Ranger* Disaster Fund, as advertized in *the Evening Telegram* were: "The *Ocean Ranger* Disaster Fund, established under the initiative and sponsorship of Canadian Newspapers Limited, publishers of the daily newspaper, the Evening Telegram, has as its main objective the raising of funds to offset the financial needs and requirements of the dependents of the 84 men who lost their lives as a result of the sinking of the *Ocean Ranger*, an offshore drilling rig, on the Hibernia oilfield on the Grand Banks of Newfoundland on February 15, 1982" *Evening Telegram* 1982m.
2. Families in the Claimants Committee received the following letter from U.S. legal counsel on October 18:

   Dear Mrs Parsons:

   On Saturday, October 15th, we had the opportunity to speak with some of the clients who are the survivors of the Newfoundland decedents, who attended the meeting of the Ocean Ranger Families Foundation.

   We reported that since our first hearing on April 6, 1983, we have amended our complaint in New Orleans and joined the American Bureau of Shipping (certificate of seaworthiness), Mitsubishi (the manufacturer of the rig, now sued in all cases), and the United States of America (for negligence of the Coast Guard). We anticipate that defendants will implead defendants Mobil and ODECO.

   On August 17, 1983, we were granted an opportunity to reargue before federal Judge Collins the issue of the convenience of the forum in the New Orleans Federal Court. We await his decision at some time in the near future and anticipate a favourable outcome. He held earlier that the Canadian claims were within the Court's jurisdiction, but as to ODECO and Mobil it was not convenient to the Court to litigate in New Orleans. However, as to the defendant Mitsubishi, he held that jurisdiction and convenience were appropriate to New Orleans.

   In addition, we have filed a complaint against the ODECO and Mobil defendants in the State Court of Texas for Matagordo County. We feel confident that the Texas court will keep jurisdiction in these cases. The issue concerning convenience is different in the state court than it is in the federal court. We consider the court to be a favorable forum.

   Depositions (sworn testimony) of ODECO employees start shortly on the issue of liability in both American and Canadian cases in the federal court.

   We are hopeful that all of the procedural roadblocks will be out of the way

soon in the U.S. courts and, if so, we will be able to bring the cases to a conclusion by settlement or trial within one year."
(Letter to Cynthia Parsons from Speiser and Krause, October 18, 1983).

3. The chapter on "Wrongful Death" in the Canadian *Law of Damages*, explains precisely this: Recovery has been allowed for the anticipated benefit that a parent could have expected from a deceased child. In some cases, awards have been made without proof or any definite expectation of financial benefit. Other cases have insisted, in the absence of such proof, that there should be no award" (Cassels 2008: 6.220). And: "If children could reasonably have expected parental support past the age of maturity, the expectation will be compensable even though the legal obligation of support may, in fact, run from the child to the parent" (6.230).
4. Collins made at least three rulings on *Ocean Ranger* cases. One, on June 5, 1984, found that Americans could sue in American courts and Canadians should sue in Canadian courts. This is consistent with the judgment that the Canadian courts were adequate to hearing their cases. The next was May 10, 1985, when Collins dismissed claims against the U.S. government. These were claims against the Coast Guard. This decision found that the Coast Guard was working within its "discretionary function" and so considered to be a separate power from the U.S. government itself. Again, this seems consistent with other rulings. Finally, on August 14, 1985, Judge Collins ruled against plaintiffs trying to take *Ocean Ranger* claims into Texas state courts. Collins ruled that he had already ruled that Canadian cases should be heard in Canadian courts and that, therefore, this would be "relitigating" without appealing. According to Professor Martin Davies of Tulane Law School, "All [these rulings] seemed fairly orthodox applications of existing law" (email July 29, 2010). Furthermore, none of these rulings was appealed which they probably would have been if they had been obviously defective.
5. A December 7, 1983, telex from C.F. Krase to Leo Barry states: "Regarding settlement, we want all our clients to know as we have previously stated, that we ultimately expect to be able to litigate their cases in the U.S. courts but we cannot guarantee when the matter would be closed or the amount of exact recover, also that the offers being made in the dependency cases are as a general rule about one-half of what we consider to be a far [sic] settlement or minimum jury verdict that we would expect in the U.S. of course, there are exceptions to the above general statement. Assuming that the client's Canadian solicitor has concluded on balance that he wishes to recommend settlement at this time, we will not oppose such recommendation in any way. Of course, we are happy to discuss any possible counters to the offers being made based on the facts of each case.... There is no doubt that together we have brought about offers far in excess of what could be expected in Canadian litigation and no client can complain that they have not received strong, vigorous and successful representation."
6. Mario Biaggi, congressman from the Bronx, was chair of the committee that passed the amendment blocking foreign workers from access to American jurisdiction in their lawsuits against American employers. In 1988 Biaggi was convicted in two corruption trials, one of which involved receiving gifts from the owner of an insurance company.

3

BLOOD MONEY: RIGHT, RESPONSIBILITY AND STIGMA

"What will you give me and I will deliver him unto you?"

When Benton Musselwhite told me that a $25,000 settlement was "a dishonour to your brother's memory," he expressed an ancient and crucial characteristic of blood money. Whatever the amount, blood money taints as it heals. The symbolic taint of blood money is one of the most powerful forces that work to manage a potential crisis in confidence in political and economic systems: the financial settlements change family members' image in the public view. This money is legally owed and often sorely needed by bereaved families, but it is also tainted with a sense of dishonour. For many *Ocean Ranger* families, the financial settlements with the oil companies were both liberating and crushing. In *Rig* Heffernan tells the story of the daughter of one of the *Ocean Ranger*'s crew. "Tracey," as she is identified, recalls her mother raising four children under the age of twelve, not receiving any insurance money and struggling until the settlement came: "Although she wanted to pursue it further, she had her children to think of, and she agreed to the settlement.... It wasn't as tough financially, but we always felt that it was blood money" (Heffernan 2009: 161). Perry Morrison's father, Scotty, could not remember what they did with their money; talking about it made him sick and he felt "like a sucker" for settling. Patricia Hickey and Patricia Ryan pushed as far as the legal systems let them, and still they wondered if they should have continued fighting. As widows with children, Noreen O'Neill and Cynthia Parsons faced a different set of issues, with neighbours fascinated by their spending and sex lives, and lawyers advising them to beware of men who were after their money. "We were warned," one widow told me (House 1987). In both roles—as mothers of dead young men and as mothers of fatherless children—women often took the lead in pursuing their families' claims against the companies.

Financial settlements are repugnant because the people we love are not "like" anyone or anything, and they certainly do not have a cash equivalent, no matter how carefully or legally the sum is calculated. Where, then, did this practice of accepting cash for life begin? Where does the stigma or taint that accompanies this money come from? How can it integrate those exiled by violent loss back into society? This chapter offers a brief history of blood money, highlighting its peculiar double nature as a right which somehow taints those who exercise it. Of course, money is currency, an empty value-marker that stands for things of real value. But money is also symbolic, as cultural anthropologists interested in social gift exchanges have argued. Money is loaded with meaning that exceeds its exchange value. This is particularly true of money that is accepted as a settlement for wrongful death.

A liberation from responsibility for vengeance and a taint that marks both giver and receiver, blood money is at least as old as recorded Western history. Keeping and settling accounts is possible only, it seems, where vengeance is revised, rewritten in other terms—legal, political and fiscal—with all the shame that entails. We are historical beings, and one crucial component of understanding our own cultural practices is to explore them in light of the tensions of the past. Families across the centuries have asked the same questions that *Ocean Ranger* families asked themselves, though in radically different historical conditions: How much is enough? How long is long enough to keep pursuing justice in financial form? What is the right thing to do with this money? In this chapter, I trace these questions back through longstanding cultural practices in the Western tradition in a brief history of blood money and the strange ambivalence of settling on and accepting a cash payment in the wake of wrongful death.

EARLY BLOOD FEUDS AND CONFLICT RESOLUTION – "GIFTS TO MELT RAGE"

Our practice of paying cash for wrongful death is grounded in more than two thousand years of struggle over the right to use violence as a "remedy" for wrongdoing. In the ancient Mediterranean cultures at the roots of Western law, responsibility for settling scores fell to family members but always within a broader socio-legal context. In most ancient communities, the gods sanctioned the feuds and/or blood prices that freed public memory from forensic reconstructions of "what happened?" among aggrieved citizens. In ancient Hebrew culture, families were responsible for taking the life of the offender. "An eye for an eye, a tooth for a tooth" is but one element in a complex network of customs that assigned value to loss and then designated appropriate settlements.[1] A blood feud was not a loss of control but rather a way of redressing and limiting violence within established practices of exchange (Barmash 2004: 185).[2] The Mesopotamian Code of Hamurabi listed blood prices for categories of people (c. 1790 BCE),[3] though direct vengeance may have also been an option (Barmash 2004: 189).

Some of our earliest descriptions of settlement negotiations are in Homer's epic poems, *The Iliad* and *The Odyssey* (c 800 BCE). In *The Iliad*, Hephaistos, the god of technology, creates a shield to protect the hero Achilleus as he wades into the raging battle of the Trojan War to retrieve the corpse of his beloved Patroclus. The god depicts two cities on the shield, two visions of mankind's possible future: a city at peace and a city at war. These symbolic possibilities protect the hero and allow him to retrieve Patroclus's corpse for the rites that usher his shade into the underworld. Among the weddings and markets of the peaceful city on Achilleus's shield, the god shows a kind of trial:

> A crowd, then, in a market place, and there
> two men at odds over satisfaction owed
> for a murder done; one claimed that all was paid,
> and publicly declared it; his opponent
> turned the reparation down, and both
> demanded a verdict from an arbiter,
> as people clamoured in support of each
> and criers restrained the crowd. The town elders
> sat in a ring, on chairs of polished stone
> (Homer 1974, Book XVIII l 571–79)

The victim's relative refused the blood money and so the elders had a problem. We hear nothing more about that symbolic peaceful city in the poem, but it remains a hopeful example of what might follow *The Iliad*'s bloody war.

Later, the poem revisits the issue of exchange, but this time, through the enemy, the grieving family of the Trojan hero Hector. In his rage and triumph, Achilleus drags Hector's lifeless body behind his chariot while Hector's family watches from Troy's city walls. The gods helped Achilleus retrieve Patroclus's body, and they do the same, in a sense for Hector's family. Zeus commands Achilleus: "Give Hector back./Take ransom for the body" (Book XXIV l: 159–65). Hector's father, King Priam, leaves the safety of Troy's citadel to enter the Greek encampment as a supplicant, bringing "gifts to melt Achilleus' rage" for Patroclus's death (145). When Achilleus accepts these gifts the war does not end, but Zeus commands respect among enemies and suggests a possible future of conflict resolution without war.

Again, in *The Odyssey*, Homer considers the problem of retribution. This time, however, there is no suggestion of a procedural resolution to the violent cycle. In this case, collective memory itself will have to be altered through divine intervention. While Odysseus was away at the war, the young men of the region moved into his house to party and try to get his wife to remarry. These "suitors" swilled the house wine, gobbled the best food, brutalized the loyal servants and harassed Odysseus's wife and son. These are brazen insults to Zeus himself, the god of hospitality, and they are answered when Odysseus comes home and massacres the young men with righteous zeal. This exchange of "bads" can be stopped only through divine intervention. The community has to be made to forget. Athena, Odysseus's patroness and the goddess of strategic wisdom, erases the memories of the families of the executed young men. In this, she liberates the community from backward-looking retribution, and thus she also liberates Odysseus from having to rule by force.

Like gift-giving, violence was an economy. To restore normal exchange in the wake of exceptional violence, the gods themselves had to act. The only way mortals could return to normal relations was if they forgot, strategically, the most hurtful moments of their violent pasts. In these ancient cultures, exacting revenge or retrieving prizes from a killer revived the body of the collectivity: vengeance for a particular death returned life to the community as a whole (Holmes 2007: 48).[4] Whether it was exacted by payment of a blood price or by a reciprocal measure of violence, the settlement of wrongful death was atonement, an exceptional payment made to reconstitute the usual socio-economic exchanges.

The responsibility of the family to enact justice was diminished in the democratic Athens of the fifth century BCE. There was no blood-money because compensation was owed to the victim, who, obviously, could not claim it (MacDowell 1978: 110). While tragic dramas like those that make up the *Oresteia* played on anxieties about the return to a mythical archaic culture plagued by cycles of violence, the city claimed the right to punish and reassert peace. The victim was not an individual in either a heroic or a modern sense but, rather, a citizen, belonging to the city more than to the family. A murderer killed part of the *polis*: thus, he or she was so unclean that Athenians held their trials in the open air to protect themselves from the perpetrator's taint or "miasma." Athenians distinguished deliberate from less-deliberate causes of death, and their trials had three clear goals: deterrence, vengeance and purification. Instead of being a private matter between the perpetrator and the victim's family, the wrongful death of an Athenian was treated as a strike against the city itself, punishable by death or exile. Even mourning was appropriated by the city: women mourners washed the bodies but their participation in public lamentations and funeral processions was constrained by law. The city also set spending limits on funerals and monuments. Athens' war dead were memorialized in group civic funerals, marked by patriotic mass eulogies, exemplified by Pericles' funeral oration. Death has long been taken as an occasion to bolster nationalist sentiments, and the particular loss of individuals translated into an investment in the life of a community divinely destined to exceed the biological ties of family.

In the Roman Empire, central authority broadened and intensified, while at the same time individual citizens were held to ideals of civically oriented stoicism: self-denial, self-control and self-sacrifice for the good of the city. The idealized Roman family included the entire household, as a quasi-independent legal institution in which the father held almost absolute power. The Twelve Tablets of the Roman Republic (c. 450 BCE) remained the symbolic anchor of customary law for centuries and laid out basic procedures for resolving conflicts. The Roman citizen was expected to live by an ethic of self-help according to the "ways of our ancestors" (Lintott 1999: 29–34). He had the right to put anyone in his household to death, though there are indications that some fathers were penalized for killing sons unnecessarily (Harries 2007: 106; Rupke 1992: 61). Outside the family, emperors struggled to sustain a "monopoly on punishment" (Rupke 1992: 65).[5] Between the ethos of self-help and the absence of policing, violence was ubiquitous and mob violence was clearly "part of how rulers and ruled communicated with each other" (Harries 2007: 116).

Why, then, did the Roman ruling class see themselves as people of the law? As republic became empire, the law was seen to express the *fides* or faith of the citizens: contracts and laws were promises grounded in a culture of good faith. Roman laws had magical power (Meyer 2004: 294).[6] The format of the emperor's edicts indicated the kind of power the tablet had: stone was more important than wood, wood more important than parchment (Meyer 2004: 2). Nonetheless, even at the end of the Roman Empire, the ethic of self-help continued to disturb central authority. As late as the fourth century CE, Constantine (or his draftsperson) wrote: "beatings (*verbera*) and murders (*caedes*) are often found to have been perpetrated when some attempt violence and others angrily resist them" (Constantine, in Harries 2007:110).

Managing memory was a different task in the ancient world than it is for us, obsessed as we are with preserving detailed archives of loss. To be remembered in the ancient world was a kind of divinity, an exclusive privilege, directly linked to the life of Rome itself, "the eternal city." Tacitus ends the biography of his father-in-law with these words: "For oblivion will bury many of the men of old as if they had been without renown or prestige: Agricola will serve for posterity, his story told and handed on'" (Flower 2006: 2–3).[7] Whereas we try to "name" our culprits, to keep them alive in memory as cautionary figures, the ancients deployed "memory sanctions." Particularly hated public figures were relegated to oblivion, their names literally chiselled out of monuments and tablets.

MEDIEVAL FEUDS AND BLOOD PRICES – "THIRTY PIECES OF SILVER"

The stigma that *Ocean Ranger* families felt when they accepted cash in settlement for death echoes the medieval ambivalence towards money. Judas betrayed Christ to the Roman authorities in exchange for money: "'What will you give me, and I will deliver him to you?' And they covenanted with him for thirty pieces of silver" (Matthew 26: 14–15). Money was the vehicle both of charity and of deepest treachery. It was inherently fraudulent, in a sense, because it always stood for something absent: money always represents something other than itself.

Excess pervaded medieval theology particularly in its Platonist metaphors of creation as light freely emanating, a gift from a self-complete, perfect creator. Medieval culture celebrated gold, silver and gems as symbols of God's gift, and yet riches were also taken to be the heart of human corruption. The medieval period set the stage for a differentiation between wholesome and reviled money-making. Anyone who handled money was tainted; the most notable examples being prostitutes, actors and merchants (LeGoff 1990). At the same time, the ideal monk was a wanderer who survived by the charity of others. Money given freely was clean. Money given under coercion, in amounts greater than necessary was unclean.

The intense symbolic importance of money was expressed in the medieval hatred of usury. As profit earned from lending money, usury was linked to sloth: the money-lender was a parasite on the real worker, feeding on his excessive desire for worldly goods. Even worse, the usurer bought into a kind of cosmic fraud: he took the symbol of plenty for plenty itself; he worshiped the empty marker rather than God's gift itself. A usurer abused the ties of reciprocity that bound the community together: "One calls anything whatsoever usury and surplus if one has collected more than one has given" (St Jerome, in LeGoff 1988: 26). Charity was legitimate excess, usury was its perversion. Usury was a theft from the community and from God, it perverted human relations of reciprocity by sinfully benefitting from others' acceptance that a "just price" or a "just wage" was expressed in market values. For the medievals, bankers and loan sharks were much worse than adulterers. At least illicit sex was love of another of God's creations. Breeding money from money was both a sterile abomination of nature and the abuse of community ties: "Making coins give birth to more coins, causing money to work without pause, in defiance of the natural laws concerning money that have been fixed by God, is that not a sin against nature?" (LeGoff 1988: 31). The great late medieval poet Dante places usurers on a par with sodomites in the *Inferno*. (Dante places some homosexuals in *Purgatory* because it is not anything like what we would call gay love that damns those in the *Inferno*. Those in the *Inferno* are damned because they invested their being in the worship of a good that only mimics God's work of creation). Usurers stole property, time and creativity, all of which properly belonged to God. As a thirteenth-century manuscript explains:

> Usurers sin against nature by wanting to make money give birth to

> money, as a horse gives birth to a horse, or a mule to a mule. Usurers are in addition thieves, for they sell time that does not belong to them, and selling someone else's property, despite its owner, is theft. In addition, since they sell nothing other than the expectation of money, that is to say, time, they sell days and nights. But the day is the time of clarity, and the night is the time for repose. It is, therefore, not just for them to receive eternal light and eternal rest. (in LeGoff 1988: 40–41)

Our own stigmatizing of blood money shares some of usury's medieval taint: blood money is a compensation for lost time (life-time), it is gained without work by those who receive it, and it is calculated mysteriously with opaque gestures towards "replacement of lost earnings," compensation for loss of "care, guidance and companionship," precedence and strategy. In medieval Europe, the right to punish, like the right to judge and to bear witness, rested ultimately with God himself. How this translated into life in this world was far from clear. Christ demanded that we abandon family as our fundamental tie, yet feud culture continued across the West well into the modern period.

Pre-Christian custom in Germanic and Celtic lands included both blood feuds and blood prices. In 600, King Ethelbert laid out a fixed set of values for lost body parts and blood prices in the event of wrongful death: a commoner was worth one hundred cows and a nobleman was worth three hundred (Berman 1978: 558). A 690 law from Wessex stated: "If a Welsh slave slays an Englishman, his owner shall hand him over to the dead man's lord or kinsmen, or purchase his life for 60 shillings. If, however, the lord will not pay this price for him, he must liberate him; afterwards his kinsmen must pay the *wergild*, if he has free kindred; if he has not, then his enemies may deal with him" (in Daly and Wilson 1988: 241–42). Payments redeemed the family's honour: they were not about "compensation" in any market sense (Berman 1978: 559). As early as the ninth century in England all freemen swore an oath of "fidelity," which bound them to keep the peace, raise a "hue and cry" if they witnessed a crime, hand suspects over to legal authorities or even pursue suspects in the king's name. Trials were held as community meetings, or "moots," centred on ordeals: if he was innocent, the accused would heal after walking across burning coals or he would sink when dropped in water or scald when dipped in boiling water.

English law adopted a longstanding Scottish practice of condemning objects or animals that caused harm, as "deodands." A deodand was "chattel" that a jury determined to have started a chain of disastrous events.[8] The deodand was either seized or its value was expressed in monetary terms. The owner was then responsible for paying a fine of the deodand's value. The bizarre-seeming trials of wells, wagons, pigs, and all manner of property can be seen as an early form of negligence fine. The term "deodand" comes from the Latin: that which is to be given to God. The payment was a kind of rationally-priced sacrifice. As Cowell (1701) put it:

> Deodand is a thing given, or rather forfeited as it were, to God for the pacification of his wrath, in case of misadventure whereby any Christian man cometh to a violent end, without the fault of any reasonable creature.

By the thirteenth and fourteenth centuries, the deodand was one of the ways juries were able to adapt the law to meet contingent circumstances (Pervukhin 2005: 239).

As late medieval customs gave way to early modern practices, law and feud cycles across Europe were a messy combination of leftover Roman practices and the customs of local clans (Nichols 1985: 141). As late as the fourteenth century in Ghent, for instance, "atonement books" record ongoing exchanges of bloodshed and money (Nichols 1985: 201–202). Starting in the eleventh century, the English distinguished punishment inflicted by the king from vengeance perpetrated by family and friends. England collectivized injury under the monarch's exclusive right to punish more quickly than other northern European countries. Legal historians have traditionally presented this early English development of criminal law as a sudden break, when feud cycles were stomped out by kings claiming exclusive right to exercise violence. In Pollock and Maitland's terms, this happened with "marvellous suddenness" (1898, in Berman 1978: 553). Recently, social historians have found that feud cycles in fact co-existed with the developing distinction between private and public justice. A king needed the political "friendship" of nobles to maintain his station by appropriate responses to the shame and humiliation of injuries (Hyams 2003: 11). Kinship and friendship networks—kith and kin—were essential to political power and to individuals seeking redress for wrongs (Neville 2010: 193).[9] Practices differed between regions and among classes, but kith and kin were entitled to violence if blood money was not paid. At the same time, families and friends had a moral and legal obligation to accept a reasonable payment: "Failure to accept a reasonable offer of compensation of blood-feud was a sin requiring penance under ecclesiastical law" (Berman 1978: 583). The emphasis was on settling rather than feuding or litigating.[10]

Late medieval (or early modern) Englishmen had a range of options following wrongful death: they could go to court, they could negotiate a blood price, or they could pursue private vengeance. Injury to one's close ally or family was a source of shame, and the need to seek redress was humiliating. In the twelfth century, "sensible men pondered the options with their friends at length before reaching for their spears" (Hyams 2003: 109).[11] Prices were negotiated but within a broad range of acceptable amounts (Hyams 2003: 88–89).[12] This idea of a given range of acceptable payments resonates with the *Ocean Ranger* families' sense there was a right time to settle. Perhaps just such a culural sense of appropriate measure is still at work in contemporary negotiations. The two Patricias—mothers of young men killed in the disaster—who pressed on into American jurisdiction faced disapproval from some of their neighbours for pushing beyond what was felt to be an acceptable range of payment for wrongful death.

Between the twelfth and fifteenth centuries, medieval men and women changed their views of serious wrongdoing. In the central Middle Ages, "homicide, arson, and theft were private wrongs that entitled the victim or his kin to demand compensation or to engage in a blood feud" (Olson 2006/7: 68). By Shakespeare's time, the right and responsibility to punish belonged to the monarchy. Medieval calls to settle amicably or to limit violence to a culturally regulated feud within a hierarchical assignment of social "worth" were replaced by a concentration of power in the sovereign.[13] This may have systematized violence by focusing it: it was a more rational justice when it worked, and a more oppressive one when it supported the powerful over the weak.

MODERN LAW – "A POUND OF FLESH"

The Protestant Reformation normalized usury and profit-making, transformed the Christian's relation to secular law and deepened the court's reach into the soul of the plaintiff and defendant (Musson 2001: 15; Weber 2002).[14] Negligence—the crucial charge in accident torts—arose in the late medieval period as part of shifts in understandings of risk and responsibility.[15] The first indications of accidental death tied to negligence emerged with sports, particularly archery, when wealthy guests were the casualties of playful thoughtlessness.

By Shakespeare's time, a merchant was still suspect, but we also see the rise of the noble, virtuous, community-minded businessman alongside of laws to support and control England's poor. Wealth and poverty became signs of divine judgment. Some fortunes were the sign of divine election just as others were proof of cupidity and thus of damnation. Similarly, some poverty was noble and pure if it was clean and hard working, while other forms of poverty were dirty, dangerous and criminal. Shakespeare's *The Merchant of Venice* is a good example of the contrast between wholesome and perverse profit-making. Antonio, the merchant, is infamously mirrored by Shylock, the evil Jew who breeds barren metals and seeks to profit from the betrayal of a good man, in a modern version of Judas's betrayal of Christ. Shylock aims to benefit from the free market exchanges of others without risking anything. He loans at interest, protected by the laws of a city he would love to see destroyed. The language of risk and gambling pervades Shakespeare's play. Resolution lies in affirming the very grounds of agreements in a culture that holds contracts in good faith. Shylock acted in bad faith because all along he hoped to kill Antonio out of vengeance. Antonio also demonstrated a kind of bad faith because he gave excessive gifts to Bassanio to bind him through a massive debt of friendship. Portia's "mercy" claimed the exceptional payment from Shylock that exceeded both the blood price (which Shylock was willing to pay, albeit under duress) and capital punishment. Shylock paid with his very identity: he was forced to convert to Christianity.

In the two and a half centuries between Shakespeare and Adam Smith's 1776 *The Wealth of Nations*, the providential order was monetarized. For the Christian medieval, "freedom" ideally meant the ability to reform one's will, to bring it into concert with the divine music of the cosmos. For the modern political economist, freedom meant a pursuit of personal self-interest, now individually defined. Smith's famous "hidden hand" saw the pursuit of narrow self-interest as distributing the benefits of profit inadvertently through the whole society. Of course, Smith recognized the competition-killing potential of monopolies: he would have been astonished by the mammoth corporations and their capacity to control a resource as geographically widespread and socially essential as petroleum. Smith, like John Locke, pictured the free market as a competition among relatively equal agents on a level playing field. The concept of a tort between an individual family and an organization like Mobil Oil would have been incomprehensible to both Adam Smith and John Locke. The image of Thomas Jefferson's yeoman farmers disputing claims before a jury of their peers still haunts American views of tort, insofar as they presuppose the parties are roughly equal before the law. Even at the very height of enlightenment thinking, Immanuel Kant recognized the problem with translating loss of life into dollar terms. He says, "everything has either a *price* or a *dignity*. What has a price can be replaced by something else as its *equivalent*; what on the other hand is raised above all price and therefore admits of no equivalent has a dignity" (Kant 1998: 42 [4: 434]).

LEGITIMATION AND INDUSTRIAL ACCIDENTS – "TO GO TO LAW FOR REDRESS IS TO CONFESS PUBLICLY THAT YOU HAVE BEEN WRONGED"

For Karl Marx, the conflict inherent in the labour market was contained by the everyday exchange of a kind of blood money: wages. In Marx's view, workers are compensated for their time, creativity and creations by receiving their market value as determined by the supply and demand of labour power. Surplus value is produced by the capitalist who buys labour power as a commodity and then uses it to produce surplus. "The surplus belongs to the capitalist; it never belonged to anyone else" as one commentator explains and then quotes Marx, "'The circumstance is peculiar good fortune for the buyer [of labour power], but no wrong or injustice (*Unrect*) to the seller'" (in Wood 2004: 136). The surplus was taken by employers and turned into capital. Reinvesting profits into what we would call "research and development" was essential for any capital holder to stay competitive with neighbouring factories. For Marx, the only true redress for workplace death would be the working class becoming conscious of themselves as agents of history. The workers slaughtered in the name of capital, be that over the course of a lifetime or suddenly in a disaster, would be avenged and redeemed, in a sense, by demonstrating to their surviving colleagues the structural fact that industry fed like a vampire off the blood of its workforce. Blood prices paid to widows and children could only be window dressing, obscuring the capitalists' defining activity, which was to turn their workers' lives into measurable and marketable things. Money, for Marx, is a condensation of human life. For what is capital, other than the time and creativity of workers?

As long as there has been capitalism there have been public policies to manage the poor, to mitigate tensions between workers and profit-makers so they do not develop into open conflict and to recover apparent legitimacy once conflict becomes visible (Green 1982). The modern age is characterized by the state assuming the monopoly on exercising violence, even in the wake of wrongdoing between individuals. But in the modern world, going to court to seek redress continued to carry a stigma that stemmed from the initial shame of the injury and one's inability to negotiate a settlement privately. As U.S. President Andrew Jackson put it: "To go to law for redress is to confess publicly that you have been wronged and the demonstration of your vulnerability places your honour in jeopardy, a jeopardy from which the 'satisfaction' of legal compensation in the hands of secular authority hardly redeems it" (in Daly and Wilson 1988: 246).

"Accidental death," with its confused lines of causation and shared responsibilities between industrialists, "imprudent" workers and government regulators, emerged with the industrial revolution. "Accident," as moderns understand it, was first and foremost an industrial way to die: it happened to coal miners and railway and factory workers. Our current range of payments for wrongful death emerged in the 1830s as the industrial revolution charged across North America, bringing death by rail and automobile with it. In the United States in particular there developed "the industrial-accident crisis" (Witt 2004: 20). Some might think that workers' compensation, private insurance and torts were competing schemes. In fact, they all developed around the same time and responded to aspects of redress that were inherited from earlier periods but recast in the modern sensibility about risk and its management.

The infamously callous cost-benefit analyses involving loss of life during the construction of large industry is an exemplary expression of this. Human life was but one cost among many in commercial, scientific and technological progress. Danger and blame at work were now categorized in administrative terms and with an increasing intent to control workers psychologically (Tucker 1990; Dodd 2006). Tort changed, too. It shifted from being primarily a tool to punish wrongs the criminal law did not, to a form of "compensation," that is, as an equivalence, a way of calculating the replacement value of a dead wage-earner.[16]

The moral stigma of "blood money," developed as the category of "accidental death," extended its reach from miners and railway workers to children in industrial towns and cities (Zelizer 1994). And as uncertainty over the purpose of the settlement grew, blood money became ever more fraught. The stigma, and its particularly gendered character, intensified when children were no longer considered wage-earners and yet their "value" in settlements grew. For the widow and mother of bereaved children, seeking a settlement was a maternal responsibility. But for the mother of a dead child, accepting payment seemed simply wrong (Zelizer 1994). Children's lives were re-valued when the insurance industry encouraged policies to cover children at the turn of the nineteenth century. Children were removed from the market as workers but they were reinvested with immeasurable sentimental worth.

In the late twentieth century, as we saw with the *Ocean Ranger* lawsuits, new forms of legal action developed to respond to the massive expansion of risks assumed by technologically developed industries.

In a 1986 ruling, Chief Justice Hickman explains the rationale of Newfoundland's *Ocean Ranger*-driven amendment to the *Workers' Compensation Act* that allowed families to receive benefits and to pursue lawsuits. Hickman writes:

> The first Workers' Compensation Act was enacted in the United Kingdom in 1897 and was replaced in that country in 1925 when Parliament at Westminster passed a new Workers' Compensation Act. The effect of such legislation was to impose absolute liability upon an employer for injuries sustained by an employee in the course of his employment and to fix the amount recoverable by an employee from his employer in such cases. This legislation had the effect of compelling an employer to obtain insurance against such risk or to be a self-insurer.... In the United Kingdom the right of tort action has been retained while at the same time conferring upon workmen or their dependents... the absolute right of recovery of established workmen's compensation benefits. (*Piercy Estate v General Bakeries Ltd.* 1986)

The relatively mechanical disasters of oil rig losses like the *Ocean Ranger* and the *Piper Alpha* were dwarfed by the massive destruction wreaked by disasters like Union Carbide's gas leak in Bhopal, India, and the more systemic destruction of entire communities by heavy metal pollution (Japan, Northern Ontario, etc). The massive destruction perpetrated by industry's mismanagement of risk led to a populist questioning of expertise at an unprecedented level (Beck 1992).

Complex social wrongs began to find voice through tort and class action suits, too. That Justice Hickman saw the Donald Marshall Jr case as controversial "in the true sense" suggests to me Hickman's awareness that understandings of responsibility were shifting and the public was demanding that legal judgment seek new paths for tracking culpability through complex decision-making webs. The question of compensation arose in Marshall's case, of course: What it could possibly mean to assign a dollar figure to this young man's eleven years of incarceration, continuous caricature by the media as a young Aboriginal thug and the lingering suspicion of guilt for a crime he did not commit?

BLOOD MONEY AND CLOSURE

Blood money is just one way in which the unspeakable betrayal of trauma is made communicable. Negotiating and settling on a cash payment is just one strand in the web of a disaster's aftermath. There has never been a time when wrongs were thought to occur simply between two independent parties, the kind C.B. MacPherson calls "possessive individualists." Though the legal language of tort suggests this, practices of fixing and exchanging blood prices have always coincided with myriad other social processes by which a perpetrator was required to confess, to seek forgiveness and to contribute to social improvement, while the injured party was required to lay the claim to rest in a timely fashion (either through appropriate reciprocal violence or settlement).

Blood money's stigma is very old. Ambivalence about blood money is rooted in cultural practices and understandings of justice that reach back to the very earliest moments of Western culture. Pursuit of a tort action is a modern version of a very ancient and enduring principle: a family's responsibility for vengeance. When they settle out of court, families are forced to translate their loss into financial terms. Accepting money closes a family's right to pursue private vengeance. Exchange of blood money is not about finding a "just price" or reasonable compensation. It is about reintegrating both parties into the community of exchange, so that the suit is settled "amicably and without admitting liability" as my family's out-of-court agreement with Mobil, Schlumberger, ODECO and Mitsubishi put it.

My sarcastic big brother's death helped finance my decade-long stay in university. But, that money was always problematic. It was not "compensation" in any meaningful sense, and it was laughable to think that Mobil, ODECO, Mitsubishi and Schlumberger considered $25,000 a significant punishment. So what was this money? It was our due, yet it was somehow shameful.

In accepting money as a settlement for a wrongful death suit, we signal our return to the relations of exchange that make up our socio-economic world. The first and most fundamental principle of blood money is the cultural demand that money can—even must—be accepted by families as a way of settling their claim for justice in the wake of wrongful death. The second is that not all corpses are worth the same amount of money. Third is that the status of the survivors changes upon receipt of money. Fourth is that such payments may be related to but are fundamentally distinct from, any punishments assigned to an individual actor. As we saw, the pressure that *Ocean Ranger* family members felt to settle is a longstanding cultural expectation that victims of wrongful death will accept a reasonable sum as a blood price. Part of what made such settlements so painful was the seemingly arbitrary distinctions that were made, the varying values placed on the lives of men who were considered invaluable by their families and friends. Once family members received their financial settlements, they often stepped back from lobbying for political reform, in part because tolerance for victims as public figures dwindles once blood money has changed hands. Though legally distinct from criminal charges, the family members' "settlements" also shift public pressure away from seeking and charging those responsible for the injury. The partial justice of the civil settlement helps the other processes recover when they are unable to hold culprits accountable. Mobil and ODECO never admitted responsibility, and yet public opinion turned against the pursuit of any of their agents. The separation between criminal and civil cases leads to such perplexing outcomes as that of the O.J. Simpson trial, wherein a culprit is found to owe reparations to the families of the dead, even though he was not found guilty of a crime. Settlements following industrial disasters are exactly like this: criminal law cannot prove guilt, and yet civil law can demonstrate some level of responsibility for the family's loss. This depends, of course, on a trade-off between severity of punishment and procedural requirements for proof of guilt.

My interest in blood feuds and the rule of law stems, in part, from my own violent impulses. Even now, when I see Mobil's logo, I am reminded that the Mobil shore manager, Merv Graham, knew that tying a line to that lifeboat could capsize it. He testified at the Royal Commission inquiry that he had informed the standby vessels. The men on the standby vessels testified that they had never heard of this danger. While Merv Graham was on shore that night, the men on those supply boats faced the raging storm with neither training nor adequate equipment, risking their lives first to try to save some of the men and then to recover their bodies. Western law rests on a distinction between public law, where the state holds the right to punish, and private law, where individuals retain a right to financial payment for wrongs done them by others. I cannot just go and kill the people responsible for the deaths of my relatives, in part because I cannot tell who they are in the midst of complex webs of decision-making, but more importantly because, since the twelfth century, the illegitimacy of precisely that kind of violence has been the essence of public peace. That said, we all still retain the right to a blood price even—and maybe especially—when criminal law fails.

NOTES

1. Leviticus 24:19–21, Deuteronomy 19:16–19:21 If a false witness rise up against any man to testify against him *that which is* wrong; then both the men, between whom the controversy *is*, shall stand before the LORD, before the priests and the judges, which shall be in those days; and the judges shall make diligent inquisition: and, behold, *if* the witness be a false witness, *and* hath testified falsely against his brother; then shall ye do unto him, as he had thought to have done unto his brother: so shalt thou put the evil away from among you. And those which remain shall hear, and fear, and shall henceforth commit no more any such evil among you. And thine eye shall not pity; *but* life *shall go* for life, eye for eye, tooth for tooth, Ex. 21.23-25 · Lev. 24.19, 20 · Mt. 5.38 hand for hand, foot for foot.
2. Barmash describes blood feud as "a legal mechanism that both assures the redress of wrongs and controls the violence to a level tolerable in a community. Blood feud is not a paroxysm of rage, careening out of control" (Barmash 2004: 185). Hyams gives a detailed definition of feud:

   Feud starts as an effort to avenge injury, generally violent injury, and often a killing.
   It represents the injury as the act of an enemy and signals a lasting enmity between the man (or men) who inflicted it and the "victim."
   The wrong that provokes and justifies feud is understood to affect a larger group that included the original victim but was in part known and even recruited in advance of trouble. Its solidarity has been set in doubt and may need reassertion.
   Given a similar sense of the vicarious liability of the injuring party's associates, they were sometimes targeted for vengeance in the principal's instead.
   The level of response is constrained by a notion of rough equivalence, requiring the keeping of a "score."
   Emotions both fuel the response and determine its quantum and nature.
   The response is ritualized in various ways to proclaim the acts to all as legitimate vengeance.
   Action from the side of the "victim" nevertheless raises the high probability of further tit-for-tat from their enemies.

> To dispel this and offer hopes of an end to the violence, something much more than the punishment of individual offenders is necessary, amounting to a veritable peace settlement between the wider groups involved (Hyams 2003: 9).

3. "The members of the victim's family participated in the process insofar as they had the right to make a claim on the slayer, but there does not seem to be any apprehension generated by the possibility of a blood avenger waiting to strike down the killer. By contrast, blood feud operated in biblical law, and places of sanctuary were needed to protect the killer" (Barmash 2004: 184).
4. Holmes' footnote reads: Wilson 27 argues on analogy with Semitic evidence that when life is paid for life, "the group's loss of blood (life) has been brought back, at least metaphorically, by a corresponding loss of blood, even if the dead man himself is not recovered."
5. "The close relationship of religion and social structure, of law and custom, is demonstrated by the *sacer-esto* formula, a curse declaring someone outlawed. This formula is tied to crimes that affect fundamental social relationships: a husband selling his wife (Plut. *Rom.*22); a child beating his parents or parents-in-law (Fest. 260. 7-11 L); a patron harming his client (Serv. Aen. 6.609); a neighbor digging up a boundary post (Dion. H. 2.74.3); and —beyond the time of the Twelve Tables —a patrician directly or indirectly eliminating a tribune of the people (Liv. 3.55). The society acknowledges how serious such crimes are. It declares the delinquent *sacer,* i.e. property of a god in terms of sacral law. As it is practiced in animal sacrifice, the property might be transferred by killing. Everybody is allowed to kill the *homo sacer* without fear of punishment. However, there is no positive legal qualification for such an act *(fas).* It is even excluded by the wording of the regulation. Any adhortation of private killing or official prosecution is lacking. Thus it is difficult to build a theory of sacrifice as the original form of execution on this evidence. Normally, the offender would not be killed. In all the cases the harmed one is in an inferior position: wife; old parents; client; a small neighbor, threatened by hunger for land of a person with power; the plebs, whose representative had been murdered. In such conflicts only, the definition of the norm would be comprehensible if we keep in mind that the Roman legal system is based on self-help. The laws are intended to guarantee a sort of social security. They are valid even if they could not be enforced. Powerful persons are integrated by Roman society, they are appealed to, but not fought. Judging by the restrictive use of capital punishment by the Twelve Tables, the discrepancy between the seriousness of the offense and the punishment in these cases is high. It demonstrates that the killing by the harmed one (and nobody else would be interested) is not really intended. The legal trick of the *sacer esto,* i.e. to define a massive, automatically valid punishment, enables society to express fundamental values even where law is suppressed by power and to do this without risking civil war" (Rupke 1992: 68)
6. "What once magico-religious authority had established, what the fides of individuals had once fortified, what pragmatic imperatives had once embraced, now

universal acceptance of the authority of the emperor was thought adequate to uphold" (Meyer 2004: 5).

7. "In Roman thought, memory was not taken for granted as a natural state or product. Rather, oblivion was considered the more normal condition, as the past receded from the present and was simply no longer connected to it. Hence, as a carefully cultivated and deliberately invoked culture of commemoration, Roman memory (*memoria*) was designed precisely in opposition to the vast oblivion into which most of the past was conceived as having already receded" (Flower 2006: 2–3).
8. For example, "Where a woman fell into a tub of boiling water, the tub is deodand" (Jeudwire 1917: 33).
9. "Incidents of feud in later medieval Scotland generated bitter enmity not only between an offender and his victim, but also between the kindred of both parties. Extant documents therefore accord friends almost as prominent a role in the re-establishment of good relations after an open animosity as they do the local magnates and heads of kindred involved" (Neville 2010: 193).
10. "The first book ever written in the West to describe a secular legal order, a manual of the early twelfth century called Laws of Henry the First (*Leges Henrici Primi),* followed the tradition of the Penitentials in emphasizing that English law prefers friendly settlement (*amor* or *amicitia*) to litigation (*iudicium*). Over a century before the *Leges Henrici Primi*, a code of Ethelred had stated: 'Where a thing has two choices, love or law'—that is, composition or judgment—"and he chooses love, it shall be as binding as judgment'" (Berman 1978: 584). "There remains in the late Middle Ages a blurring of the lines between private and public wrong. Concomitantly running parallel to emergent penal policy on the continent and in England was a form of dispute resolution that focused on the accommodation between parties and on charity" (Olson 2006/7: 74).
11. "The real crux is the ability of the world-be avenger to convince first himself, then his friends, and finally neutral bystanders that his chosen solution (blood vengeance, or money compensation, or some further possibility) was right and proper. He must argue his case within accepted norms. If he does this in court, he will plead that his action is justified by the rules in force" (Hyams 2003: 6).
12. "The social groupings into which their compilers divided society for the purposes of wergild payments (for example, two, six, or twelve hundred-shillings groups), also afforded men a literal sense of their own worth, an apparently objective measure of what they might with honour accept with the killing of a close kinsman. We should probably not take the sums as fixing an automatic price. Instead, they set an acceptable range within which to seek or resist acknowledgment of the special strengths of the deceased, perhaps his reputation, his warrior skills, and wisdom as a counsellor. To compare this system of valuation to horse-trading or dealing in used cars misses the overt concentration of all parties on honour and face. All the same, the wergilds must have operated in real life as rough indices of social worth" (Hyams 2003: 88–89).
13. "The line between excusable and felonious killing proved especially permeable when jurors faced those who killed their subordinates. Some people obtained

pardons after killing in an overzealous exercise of what contemporaries deemed their legitimate authority. With social and gender relations premised upon patriarchal control, masters were expected to chastise their servants, and husbands their wives, to maintain order" (Kesselring 2003: 98).

14. "In the fourteenth century, many churchmen put forward the belief that since recourse to the legal system symbolised the collapse of normal obligations of Christian brotherhood, those who represented the human face of the law (and whose professional life in fact depended on litigation) must be stirring up animosity within society since they stood to gain by it. John Wycliffe, for example, was particularly vehement against the moral reprehensibility of formal dispute settlement" (Musson 2001: 15).
15. "People judged to have killed by accident or misfortune also received pardons *de cursu* after conviction. In 1279, Edward I told some of his judges simply to acquit such people, thus freeing them of the forfeiture attendant upon conviction and the fees associated with the pardon. No permanent change in legal practice resulted, however. Only in the late fourteenth century did courts again proffer acquittals for accidental homicide" (Kesselring 2003: 96).
16. "'That in one form or another a fair amount of negligence in the sense of doing what a reasonable man would not do, or doing what he would do, was covered by medieval law.' It was in the period from 1825 onwards which Winfield's research indicates as being the most 'fruitful' in terms of the development of an independent tort of negligence" (Winfield in Klar 2003: 147).

4

COMMEMORATIONS: GIVING SHAPE TO LOSS

"There was nothing moving out there except seagulls"

On February fifteenth, I go to the shore. Whatever the weather, I watch the waves and I open my thoughts to remembering Jim, his co-workers and others who lost family and friends in the *Ocean Ranger* disaster. The reality of Jim has faded, leaving images and impressions of the twenty-four-year-old man that was: the sarcastic and supportive book-lover, the adventurer who broke our parents in, the "smart one," the worker worth about $25,000 and the one who, along with eighty-three others, died in the process of bringing Newfoundlanders and Canadians into full maturity as offshore oil producers. My private commemoration of Jim and his co-workers has shifted more and more to political questions over the years. This might have something to do with teaching Marx on the *Ocean Ranger* anniversary for more than a decade in the King's Foundation Year Programme. Mostly, though, it comes from thinking about what my parents went through when they lost their eldest child and then "negotiated" a financial settlement with some of the biggest corporations in the world in the midst of the most devastating loss of their lives.

In this chapter, I consider the ideological importance of commemorations of the *Ocean Ranger*'s crew. From personal remembrances to official, public events, the ways we remember are as political as events themselves. This is true particularly of disasters that result when the profit drive of corporate activity goes unharnessed by government regulation. Commemorations are ambivalent by nature: they express healing and genuine closure, and at the same time they can mask a false closure that helps the world of commerce just "get on with things."

With the first news of the *Ocean Ranger* disaster, "the whole island was cloaked in grief" (Moores 2011), and a shocked country and world marked the loss by sending telexes of condolences to Mobil Oil Canada.[1] Public commemoration took many forms. Flags flew at half mast across Canada, Newfoundland Premier Brian Peckford declared a day of mourning, and thousands watched the memorial service to the *Ocean Ranger*'s crew that was broadcast from the St John's Basilica. Even on the first anniversary of the disaster, emotions ran high across the island, and churches were packed for evening services. At a service

"Anchor," a model of the Ocean Ranger *monument, by A. Stewart Montgomerie (photograph by and permission from artist A. Stewart Montgomerie, photograph courtesy of Marie Wadden)*

broadcast by CBC radio, a minister repeated the time-honoured consolation that the men did not die in vain and would not be forgotten:

> We are in such a world that every good that we enjoy, every quantum leap in man's progress, every remedial measure to make life safer and more meaningful, springs from the graves of those no longer with us. Through the story of human progress and freedom and the whole pattern of our existence, runs the red current of sacrifice, constantly being replenished by people such as we remember today. (Baker 1983)

According to this preacher's redemptive story, a noble sacrifice was made by the men who died, but further sacrifice was called for. In the coming months the families would have to sacrifice even more as their private attempts to heal were disrupted by the public revisiting of the details of the disaster in the numerous inquiries and legal struggles. The preacher encouraged families to be strong because they, too, were contributing to progress and to ensuring that this "never happens again." The Oilfield Technical Society presented a plaque with the men's names to members of the Families Foundation, along with a commitment to complete a public memorial site for the men. On March 6th, three weeks after the first anniversary of the *Ocean Ranger* disaster, the *Evening Telegram* printed a poem by one of the men who died, Greg Tiller. Written on his last trip home and found by his family after the disaster, Tiller's poem takes on prophetic—and commemorative—meaning:

"Rig"
Huge Iron Island
37 stories high, two city blocks square
impervious
to the attacks
of an indignant ocean.
Mother Earth is victimized
by this man-made mechanical rapist,
and grudgingly surrenders her treasure
to the oil magnate with the
dollar sign eyes.
Our mutton-headed people trail
behind this pied-piper, bickering
over the loose change
falling
through the holes in his pockets.
Mother Earth created us, raised us,
taught us, sheltered us, and this
is how we repay her.
Beware, she shall have her revenge.

(Greg Tiller. courtesy of Steven Porter, editor of D.C. Publishing, 2009, "Random Thoughts: Memories & Writings of Greg Tiller," compiled by Steve Porter)

Like inquiry reports and financial settlements, commemorations represent emotions and feelings. However, whereas formal legal processes work to re-establish public confidence in the capacity of social forensic processes to produce "truth," popular commemorations express a different aspect of reality. The most authentic commemorations challenge the simple narrative of shock-forensics-reform-action that I argue the inquiry report and financial settlements exemplify. In psychoanalytic terms, all re-tellings of this story repress some aspects while they work through others. The works I focus on here are *But Who Cares Now*?, Ron Hynes' song "Atlantic Blue," the monument at the Confederation Building in St John's and two major works from 2009. It was not until 2009 that two major works revised the bureaucratic account of the inquiry report, marking the end of a period of relative latency when some of the most disturbing aspects of the *Ocean Ranger* disaster were repressed, especially details of the evacuation. Both Lisa Moore's novel *February* and the creative non-fiction of Mike Heffernan's *Rig* return to the most intimate details of the disaster. Each 2009 work reclaims the men's last moments from the silence of the inquiry report, and at the same time they recast the disaster as something that happened before, in a Newfoundland that is now history.

COMMEMORATIVE ACTIVISM: *BUT WHO CARES NOW?*

The Ocean Ranger Families Foundation was formed by the end of April in 1982. Over the next few years, the work of the foundation shifted support and advocacy for the bereaved families into a more explicit "commemorative activism" (Newhook n.d.; Verberg and Davis 2011) that focused on health and safety in the offshore. Both federal and provincial governments invested in the group support the foundation offered to the families in the early days with a $75,000 infusion of funds. As the foundation's executive administrator tells it, that initial support came when he approached former premier Joey Smallwood, who in turn telephoned federal Minister of Energy, Mines and Resources Jean Chrétien to secure the funding (Newhook 2008). The first major publication commemorating the *Ocean Ranger*'s crew was *But Who Cares Now?*, a book the Ocean Ranger Families Foundation published in cooperation with Memorial University sociologist Dr Douglas House. This was "action research," where the scholar teamed up with "subjects" as partners in research and writing. Some family members, trained by Dr House conducted interviews and worked on the manuscript. In psychoanalytic terms, this is a classic work of mourning, whereby the inexpressible loss is represented in stories and, in this case, turned into an object—a bound and published book—that expressed the family members' perspective both as memoirs and as university-validated research. *But Who Cares Now?* politicized the families' experiences; it is "commemorative activism" in that it showed the family members as political actors, and it challenged the bureaucratizing narrative of the inquiry report by urgently asking, first, is this economic development worth the human cost? And further, what will it take to keep governments focused on their regulatory responsibility?

MEMORIAL SERVICE AND MONUMENT

For many of the people who were most directly touched by the disaster, the ordinary physical world was reinscribed with the symbolism of loss. The landscape itself took on symbolic importance: trees planted by the dead, buildings built, anything touched by the hand of the lost man stood out in new ways. The trees Jim planted changed my parent's field. Even the Atlantic Ocean became a sublime monument to the dead, especially in its most ferocious moods. Mundane objects became memorabilia. The cannons that were recovered in a dive by Samantha Gerbeau's father still mark the drive up Signal Hill in St John's. One widow was sent a belt buckle from a one-time *Ocean Ranger* worker. It is an etching of the rig with "1976–1982" inscribed on it as a memorial to the rig, the men who died and as a keepsake, presumably, for those workers on other shifts who survived the "Ocean Danger." The publisher for the oil industry magazine the *Alberta Report* argued that the first Mobil Oil annual report after the disaster should be "two full pages of 84 names and photographs, and above them we will see a line that reads, 'These gave their lives that mankind could find oil beneath the bed of the seas.' That seems excessive. It happens to be true" (*Alberta Report* 1982). Even the wreck of the $120 million "Cadillac" of the oil drilling fleet has monumental value. The diving team who searched the wreck in the days after it sank held a service and cast a wreath over the wreck, and another small service was held when the rig was sunk into its final resting place in "a designated dumping spot," where it does not pose a threat to commercial sea traffic (*Evening Telegram* 1983). Songs, poems, a play, a flood of newspaper and magazine articles, numerous chapters in popular collections of marine disasters, television docudramas, memorial services and sites, all spin threads of the *Ocean Ranger* story.

Every year, on or near February 15th, the students of Gonzaga High School offer a memorial service for the men who died on the *Ocean Ranger*. Originally a service dedicated to the four graduates from the school who died that night, the organizing committee was asked by family members of other men to embrace all the *Ocean Ranger* dead. Students, now born well after the disaster, light a candle for each of the men, and a speaker is chosen to deliver a reflection on the meaning of the loss to the school. Such formal commemorations shore up the community's sense of accomplishment in the wake of disaster. Not only have we not forgotten, but we collectively commit to ongoing mindfulness. In the 2011 service, Noreen O'Neill delivered the following thank-you on behalf of the families:

> Twenty-nine years ago today my son was nine months old; I became a widow and single parent over night. My husband, my loved one, was lost at sea. We were lost without him. I could never say he was dead, he was just gone away. Family, friends and the community helped us during those difficult times. And we the *Ocean Ranger* family members came together and cried on each other's shoulders. Today many of our children and family are working in the oil industry on land or at sea. We remember why we lost our loved ones, we will never forget, we hope and pray the oil industry and government will never forget and always keep safety first.
>
> I am very grateful that you, the students and Gonzaga community, are taking time to reflect on the loss of the men on the *Ocean Ranger*. I want to thank you on behalf of all eighty-four families, especially the families who cannot attend. Twenty-nine years ago you, the Gonzaga

> students of today, were not yet born. We thank you for this tribute.
>
> We take hope in you remembering; that men and woman who work off shore today stay safer because you refuse to forget. (O'Neill 2011)

At the Gonzaga High School memorial service in 2009, students sang Ron Hynes' "Atlantic Blue," the most popular of a number of commemorative songs. An extensive reading of Mike Heffernan's *Rig* reclaimed the disaster even further from the administrative authority of "the" *Ocean Ranger* story given by the inquiry report. The church was packed as members of the Canada–Newfoundland and Labrador Offshore Petroleum Board, the oil industry, politicians, family and friends of the lost men all gathered in remembrance. There was no mention of the responsible companies, nor of the failure of government regulation, though there was considerable mention of the importance of vigilance around safety.

After the service, there is a wreath-laying at the memorial sculpture. The memorial sculpture was built in 1985 on the grounds of the Confederation Building, the home of the Newfoundland and Labrador House of Assembly. It was erected by the Oilfield Technical Society, a club dedicated to creating a community for oil industry newcomers to Newfoundland. Some family members would have preferred a site on Signal Hill, the symbolic centre of St John's, where historically messages were telegraphed across the world. In the end, the committee accepted the grounds of the Confederation Building (Bursey 2010). Initially the Monument Committee hoped to place a "simple and tasteful" monument at "Ladies Lookout" on Signal Hill, "because this spot is traditionally the place where women watched for their men to return from the sea," as the committee chair explained (*Evening Telegram* 1982af). Robert Strong, then president of the Oilfield Technical Society explains:

> Choosing the proper location became a serious problem. Several of us thought Signal Hill would be appropriate because it overlooked the ocean and stood for security. Parks Canada wouldn't hear of it. We pulled as many strings as we could—I even went to Ottawa to plead our case—but they just wouldn't hear of it. I thought, *Jesus, fifty-six Newfoundlanders went down on that rig and we can't put something up there to honour them because Parks Canada, not Parks Newfoundland, says so?* Everyone was disgusted with the whole thing. (in Heffernan 2009: 191).

The *Ocean Ranger* memorial sculpture is a jagged mass that suggests rather than depicts an anchor backed by an enclosed garden. The sculpture hangs out in the open, buffeted by the wind that blows in from the harbour. A small path leads to the enclave where the men's names are engraved and mounted on an embracing wooden plank fence. Reminiscent of a patio, or even the planks of a small boat, the landscaped enclave is a hospitable place in contrast with the sculpture's exposure. The space was created as a symbolic meeting place where the men are memorialized as lost *Ocean Ranger* crewmen, but also reclaimed in a sense by the people of Newfoundland.

At the monument's dedication in 1985, seafaring fatalism mingled with the determination to preserve the rig and its crew in memory as an ongoing commitment to safety in the offshore. As master of ceremonies, the legendary actor Gordon Pinsent said: "This memorial will stand forever as a permanent reminder of the bitter price we have paid so soon in the harvest of our new found riches under the sea. It should also serve as a constant reminder of the need to protect our marine workers." Cle Newhook from the Families Foundation vowed that the men will "never become statistics." Energy Minister Bill Marshall emphasized that "as a seafaring people, we realize the element of risk." He continued: "Words alone cannot express feelings as profound and as personal as those generated by the Ocean Ranger tragedy. In such circumstances, society often turns to its religious bases and its poetical or architectural expressions" (Marshall 1985). And federal Justice Minister John Crosbie included elements of all the reported statements: "The sea will never be tamed" and also that "Both governments will see that their recommendations are carried into effect, unless there's some terribly convincing reason why that should not happen" (*Oilweek* 1985).

"ATLANTIC BLUE": IS THAT YOU?

Ron Hynes' "Atlantic Blue" is by far the most popular of a number of songs memorializing the *Ranger*'s crew. Hynes is a Newfoundland icon whose mingling of traditional and pop music combines with a poetical clarity. Hynes' most famous song, "Sonny's Dream," is said by some to be an "anthem for Atlantic Canada" (MacKay 1989: 31). "Sonny's Dream" is a melancholic ballad where the title character is left as his mother's only solace, "Sonny don't go away. I am here all alone./And your daddy's a sailor who never comes home./And the nights get so long and the silence goes on...." The romance of life on the ocean was central to the aftermath of the *Ocean Ranger* loss. Many mainstays of the romance of fishing transferred easily: long absences from home, danger, the male-dominated, isolated, utterly artificial workplace perched in a menacing Nature. The central figure of the fishing romance, the independent fisherman working freely in his small boat, had already become like a character from a fable, thanks to the new reality of huge trawlers, factory ships and the international companies that own them.

When I accosted Hynes after a performance at the Union Street Cafe in Berwick last year to ask him about family members' responses to the song, my mother sneaked up behind me to thank Hynes from the bottom of her heart. Hynes focused on Mom immediately, and so my impromptu interview was hijacked by our performance of the answer to my question. Hynes was unsure if he should keep performing the song: maybe family members were tired of it, and maybe they wanted the song, like the disaster, put to rest. According to Joyce Dodd, the answer was a resounding "No." I am sure this has happened more than once to Hynes since he released the song in 1996.

"Atlantic Blue"
What colour is the heartache from a love lost at sea?
What shade of memory never fades but lingers to eternity?
And how dark is the light of day that sleepless eyes of mine survey?
Is that you, Atlantic Blue? My heart is as cold, as cold as you.
How is one heart chosen to never lie at peace?
How many moments remain? Is there not one of sweet release?
And who's the stranger at my door,
To haunt my dreams forever more?
Is that you, Atlantic Blue? My heart is as cold, as cold as you, as you.
I awake in the morning as the waves wash o'er the sand,
I hold my hurt at bay, while I hold the children in my hands.
And whose plea will receive no answer?
Whose cry is lost upon the wind?
Who's the voice so familiar,
That whispers my name as the night comes in?
And whose wish never fails to find my vacant heart on Valentine's?
Is that you Atlantic Blue? My heart is as cold,
My heart is as cold, my heart is as cold, as cold as you, as you, as you.

In "Atlantic Blue," Hynes dives to the heart of suffering, using colour and shade to reunite the men with those who loved them. A disintegrated whole reunites in a vague horizon of blue as Hynes gestures towards the dead men's final moments without needing to figure them explicitly. The pervasive ache of unresolved mourning is figured in the dark light surveyed by "sleepless eyes." The unburied shades of men lost at sea are gathered into the "one heart chosen to never lie at peace." The song is a series of unanswerable, melancholic questions, a yearning that is itself a form of remembrance, even as it begs for the comforts of oblivion. The only positive statements unify the living and the dead: "My heart is as cold... as you," and the woman watching the waves shift the passive sands, who wakes to "hold my hurt at bay" and to hold the children. The lyrics resist the closure of a redemptive conclusion. Instead, the song offers closure in another sense because it associates the *Ocean Ranger* deaths with the romanticized melancholy of loss at sea: "what shade of memory never fades but lingers to eternity?" Listeners are invited into a luxurious melancholy, a traditional home in loss.

Hynes' song is akin to what folklorists call a "revenant" ballad, where the dead returns to comfort but also to mystify the survivor. The singer here assumes the subject position of the woman left on the shore, a familiar character of maritime romanticism.[2] Her feelings are figured in a question: "What colour is a heartache?" The answer is another question: "Is that you Atlantic Blue?", suggesting at once the ocean's depth, unresolved mourning, uncertainty and "the queer cold sensation" that presages death (Halpert 1991: 102). The lyrics also question time: "How many moments remain? Is there not one of sweet release?" The melancholic widow is haunted by the "revenant visitor," the wraith who stops by on the night of his death to say goodbye (Halpert 1991: 100) and maybe to command a return to life, away from lingering grief (Buchan 1991: 120). "Atlantic Blue" offers no answer to the questions, leaves them unresolved in a haunting that chills the heart "Is that you?"

RIG AND *FEBRUARY*: "MEMORY AND EMOTION, NOT HISTORY"

Mike Heffernan's *Rig* and Lisa Moore's *February* each explore the hinge between personal attempts to work through and the broader public memory of the disaster. Both are free from the formal constraints of the quasi-judicial inquiry or academic historical writing. *February* conveys feeling, affect, and makes no claim to accurately report on real events. *Rig* accepts the testimony of people as they recall the *Ocean Ranger* disaster almost thirty years later. These are well-worked tales in many cases, and rather than subjecting them to the formal "objective" standards of an inquiry, oral history simply transcribes them. Their validity comes from their status as testimony of a traumatic experience. Like *February*'s literary representation of affect, *Rig* channels feeling and recognizes the complexities of traumatic memory without capturing interviewees' experiences in any overarching narrative (beyond the author's editorial work to turn the interviews into anecdotes).

The fact that it took almost three decades before the *Ocean Ranger* families' reflections in *But Who Cares Now?* (House 1987) were extended by other writers is a tribute to the mastery of Justice Hickman's inquiry report on the one hand and the silencing effect of the blood money from financial settlements on the other. Accepting blood money forced family members to give up their public claim that their particular men had a social value that exceeded any monetary value. In Marx's language, family members admitted that commodity fetishism applied to the very biological life of each man. At least in legal terms, the man's life became equivalent to a cash payment. Without any criminal sanctions, or even a court ruling on the lawsuit, the companies ultimately won the struggle over the terms of the settlement. The almost thirty years between the release and reception of Justice Hickman's report and the publication of Lisa Moore's *February* and Mike Heffernan's *Rig* were filled by a relative silence. These very different works each reclaim the *Ocean Ranger* story from redemptive narratives of technological "lessons learned" and romantic tales, where, as the traditional song goes, "men must work and women must weep" (Kingsley 1849).

The heart of Justice Hickman's Royal Commission report was the reconstruction of the events of the night of February the 14th and early morning of February the 15th from the traces the desperate men left on their equipment. The men had not known how to use the ballast control equipment, so even the slight anomaly of a broken portal in the ballast control room set off a chain of disastrous events. Hickman's report reconstructed the men's last hours trying to correct the rig's increasing tilt in the terrible winter storm. However, the report relinquished any claim to represent the men's subject positions after the final mayday call. The chaotic evacuation and violent deaths of the men were left in peace, so to speak—cloaked from the prying eye of the forensic reconstruction of the "cause" of the disaster. The capacity of the two 2009 works to return to the point of death and to represent the feelings of the shock of the disaster is, for the imagined character of "the people of Newfoundland," a move beyond the false closure imposed by the inquiry report.

Mike Heffernan's *Rig: An Oral History of the Ocean Ranger Disaster* presents the *Ocean Ranger* story as explicitly political, with individual characters, like rig supervisor Jimmy Counts, standing for the negligence and exploitation of the industry. In Hickman's inquiry report, Counts comes across as simply refusing to respect the ocean and stunned by the rig's susceptibility to its harsh environment. Counts did, after all, demand that a helicopter crew fly into that hurricane-force storm so he could see with his own eyes the empty place where the rig should be. In *Rig*, Counts is a bully, a bigot, an outsider of the worst sort who has no respect for the sea or maritime culture. Counts personifies anxiety about a slide back into a colonial past, only this time the colonizing power is American industry instead of the representatives of the queen or federal parliament.

In *Rig*, Heffernan selects and arranges oral histories. His selection and arrangement of these memory pieces echoes themes from horror, disaster and Newfoundland folktales. The book opens with a collection of excerpts to frame the interviews in an "Historical Note." The inquiry report's naming of the owners, builders and operators of the rig leads into the *Evening Telegram*'s story "Rig Goes Down: Bodies Sighted" from February 15, 1982. Marie Wadden's CBC story on the release of the report comes next: "Everyone gets some blame: the Federal and Provincial governments for having inadequate legislative control offshore, the oil companies, Mobil and ODECO, for not providing proper training and life-saving equipment for the rig's crew, and agencies like the U.S. Coast Guard and Search and Rescue. Everyone gets some blame, largely for their lack of foresight that a marine tragedy of this magnitude could ever have occurred" ("Here and Now" August 13, 1984). Finally, *Rig* quotes another prominent Newfoundlander, Rex Murphy, who said the following on *The National* on the twenty-fifth anniversary of the disaster:

> War and work cost a lot in Newfoundland. They always have. The *Ocean Ranger* disaster flashed through the circuits of these common memories connected with them, but with an additional irony. The offshore was not the seal hunt or the ancient fishery. Oil was modern. The rig was a splendour of engineering and technology. The jobs belonged to an industry that might walk us away from dependency and from those old, harsh patterns of hard times and inescapable perils. The offshore was for many Newfoundlanders all hope and future, but here we were on February 15, 1982, in the last quarter of the gleaming twentieth century about to veer into a new, more accommodating richer encounter with the sea and its resources, and that terrible bell rang once again. Families hurled into grief, communities lacerated, the whole province once again struggling to absorb an assault too large for anything but time or faith to carry. Twenty-five years on, it is, of course, still being felt. (in Heffernan 2009: 3)

Rig could have been called, "The *Ocean Ranger*: The people's inquiry"—Heffernan's epilogue shows how he, like T. Alex Hickman, travelled to the heart of the matter, amassed documentation and heard versions of the truth from the mouths of the key actors. This looking into the past is a cautionary tale, a tale of "memory and emotion, not history" Heffernan claims (2009: 200). And yet as "oral history" *Rig* also claims authenticity by virtue of the author's having "travelled among other people's pain: rig workers, victims' families, emergency responders, priests, government officials and reporters. They are all ordinary people, many of them from very different backgrounds but connected by the scar left from the longest week of their lives, talking about something so traumatic that it often defies words" (200).

For Heffernan, as for Hickman, the *Ocean Ranger* disaster is a learning story, but in this case, it is a story that warns us to beware of strangers, to mind that Newfoundlanders remember who they are when they enter into contracts, particularly with American oilmen. Heffernan's accounts do not have to meet a standard of evidence: memoirs stand for themselves; they are self-validating. In this, Heffernan offers the kind of memorial work that testifies to the reality of trauma time rather than the kind that anchors itself in truth claims built on objectively demonstrable evidence. Heffernan also reasserts the culture of storytelling as its own form of history, and so the monologues that make up his oral history do much more than simply preserve the details of individual people's experiences. These tales assert the right to tell the story, and the right to tell one's own story generally, reclaiming "the" story back from the formal legal processes of the inquiries and the lawsuits.

Part One of *Rig*, entitled "Madhouse," draws the characters of the American oilmen with their relentless focus on profit into tension with the more community-oriented Newfoundlanders. Jimmy Counts, ODECO's rig superintendent, is described as a bully and a bigot who resents hiring Newfoundlanders. He would single them out to insult them: "a dumb fucking Newfie" (Russell, in Heffernan 2009: 15). Heffernan revises the inquiry report's softening of the Americans' antagonism towards Newfoundland and maritime culture. In Heffernan's version, Counts stands for the oil industry. Generally, he is an acceptable villain, because he did not die that night. Mobil's top man, toolpusher Kent Thompson, did die and he receives gentler treatment. Counts is presented as in denial of the fact that the *Ocean Ranger* is, in fact, floating in the Atlantic Ocean. He is remembered by a few of the interviewees as claiming the rig could not sink. Following the severe list, what the inquiry report called the "prelude" to the disaster, Counts reportedly "cut in" to a discussion about safety to insist, "'Don't forget, ya'll—this rig can't sink.' He then walked out" (Russell, in Heffernan 2009: 17). A difference emerges between Newfoundlanders who knew "how to handle" the arbitrary exercise of authority by Americans, and those who did not, those who responded with their fists or by quitting their jobs. One testifies: "The Americans knew the oil business, but that's common sense and stupid to think any different. It was their rig and they'd been on it for years before they came up here. But a lot of Newfoundlanders didn't like being told how it was going to be. It didn't help matters that Jimmy'd fire two or three guys just about every shift. If you were standing around doing nothing you'd be gone. Some guys would get off the chopper and if he didn't like the looks of them they'd be sent right back home again" (Major, in Heffernan 2009: 39). Men retold a story of two Newfoundlanders fighting back against the Americans. Two brothers from Torbay are said to have beat Counts up. Counts retaliated by having them suspended by a crane over the Atlantic, dressed in light clothing. The hostility between the Americans and the locals was often explicit. Counts is remembered as having expressed his power arbitrarily, to keep all the men constantly aware that, as Kent Thompson is reported to have said, "there's sixteen thousand people looking for your job" (16).

Counts is remembered as being even more arrogant than the rest of the Americans: "Jimmy Counts, however, was another breed altogether. He thought he was God. Out there, they all thought they were the dominant species" (Crowe, in Heffernan 2009: 25). The naked exploitation and sense of entitlement of these foreign managers operating in Canadian waters with a largely Newfoundland crew finds voice through Heffernan's oral history.

> It was still very early on when ODECO's rig superintendent, Jimmy Counts, showed up. Standing next to me with his supervisor in New Orleans on the line, he said, "I'm flying out to save that rig." I thought, Don't you mean you're going out to save those boys, those men? For me, it was a telling example of the indifferent attitude of the industry. It cost $100,000 a day to keep the rig going, and it was common to charter a plane just to fly in a five dollar screw from the U.S. For them, human life seemed insignificant to the drill continually turning right. (Flynn, in Heffernan 2009: 60)

This same exercise of foreign power over Canadian citizens had been described in the inquiry report in neutralizing terms as a lack of clarity in the chain of command, or a confusion in priority between the marine captain and the toolpusher. This was more than an organizational problem. It has a culturally symbolic component relating to the, at one time, notorious authority of a sea captain; from the seventeenth century on, the "dreaded Fishing Admirals" were "judge, jury and executioner" (Earle 1998: 98). Imagine the message sent to a young man like Greg Hickey, by the incident he described to his mother: he saw Mobil's toolpusher force the rig's captain to suppress a telex report of a significant oil spill. This could only have reinforced the general sense that, whatever this workplace was, it was no ship (Hickey 1984).[3] Not only did Counts have a tension-filled relationship with Mobil's top man on the rig, he also had no training in keeping the rig stable. His testimony at the inquiry was remarkable. He received a call around 7:00 p.m. that evening to tell him about the broken porthole. He heard nothing further until 1:30 a.m., when he was informed that the rig had been "evacuated." He famously made his way through the snowy streets to jump into a helicopter at 3:00 a.m. to fly through the storm to the scene. "What did you think you could do?" he was asked. Wadden's story that night focused on Counts' hubris. She notes: "Counts kept insisting that 'If they'd had a problem, they would have called me.' But they didn't call him... even though he was the company man in charge that night. This is one of the big puzzles surrounding the disastrous events of that night" (1983).

Heffernan's "Madhouse" also introduces us to the community of the traumatized who populate this book. Many of *Rig's* monologues describe the lingering effects on people with relatively incidental relations to the dead men, including a waitress, journalists, a police officer. Some members of this community recall having to come to terms with the loss, turning away from mourning and back to active life, either with formal psychological counselling or through a sudden realization that time was too precious. Others are still trapped, to an extent, in that moment or haunted by dreams and regrets.

> I got the 'Why me?' syndrome. I thought Why wasn't I out there? How come I was so lucky and no one else was? It certainly affected how I felt inside and how I dealt with people. Sometimes it was hard to keep my composure, and I went through a long period where, for instance, if someone cut me off on the road I'd curse them down and shake my fists at them. I'd often get upset at work, too. Around six years ago, I got some help and came to terms with what had happened as best I could. It was something I had to do. (Wall, in Heffernan 2009: 48)

One of the most radical moments in *Rig* is when Heffernan parses an interview with an ODECO radio operator who deploys direct experience, the—"I was there, and I swear it was this way" perspective—to counter the stabilizing causal narrative of the inquiry report:

> A wave could never have been hard enough to break that four-inch thick glass. Never, never, never. Those hoses busted that porthole, caused water to get in on the control panel and short circuit it. That was the company's fault—it was a design flaw, just like how she couldn't recover from a severe list. I'll swear to that on my mother's grave. I'll swear, I'll swear, I'll swear. (Dyke, in Heffernan 2009: 51)

Heffernan classes this monologue, this counter-causation story, with the others in the "Madhouse" because, as the interviewee clearly indicates, he has been deemed unworthy of participating in the official discussion of what happened and who was to blame. He lists official after official who seemed to listen intently to what he had to say and then shut him down altogether.

Heffernan leaves the question hanging without ever articulating it: is there something to this?

Heffernan's collection reconstructs the "tightly knit community" addressed by the inquiry report: "Everyone knew someone out there" (Flynn, in Heffernan 2009: 61). Here, though, the community is traumatized. Everyone in the book, including the waitress at a bar frequented by oil workers, the spokesperson for Mobil Oil and first responders as well as family members suffered flashbacks, anger, nightmares, guilt and often uncertainty about whether they had done all they could in their community roles. Heffernan lists last names for all his interviewees except for relatives. "Elaine," the sister of one of the *Ocean Ranger* crew, recalls: "On one of the anniversaries, I remember the headline on the front page of the *Evening Telegram*. It talked about the cost of oil and that we've already paid too high a price for prosperity. It was certainly too high for me and my family" (145). This was true of the Mobil workers gathered to respond to the emergency; it was also virtually true of the province of Newfoundland, constructed here as a unified community traumatized by outsiders. As one worker puts it: "The companies were so caught up with getting oil out of the ground that they didn't give any thought to things going wrong, like the rig capsizing, because land rigs don't capsize, and that's where they had all come from" (Flynn, in Heffernan 2009: 61).

In Heffernan's collection, the stories of the final witnesses of the *Ocean Ranger* read like horror stories. The second mate from the *Seaforth Highlander* retells the story of the men who had miraculously launched a boat off the sinking rig in monstrous waves. For this sailor, those men did their part: they got out of their bunks and into that lifeboat, they launched it somehow into the raging storm off the sinking rig, they fired their flares, and it still was not enough. He says: "A hatch in the lifeboat opened and someone emerged. The mate and four seamen—I couldn't see who—threw a ring with a rope attached. The guy caught it and tied it to the railing of the lifeboat. The next thing, seven or eight of them came out and stood along the gunwale. Then the lifeboat just tipped over and they were all thrown in the water" (Higdon, in Heffernan 2009: 63–64).

Heffernan collects tales of hopeless heroism, of men unprepared for the uncanniness of their situation. He repeats the story given in the inquiry report of the last sightings of the rig and the dying men, but this time in unapologetically horrifying terms: A deckhand from the *Boltentor* recalls that when they arrived at the *Ocean Ranger* that night, the rig that should have been "lit up like a Christmas tree" was dark but for four or five lights, "We circled but found nothing—no survivors, no lifeboats" (Kane, in Heffernan 2009: 69). Responding to the *Seaforth Highlander*, they tried to rescue the men from the spilt lifeboat but the extreme storm blew the men and their blinking life jackets away into the waves. "Then the *Norderator* [a supply ship for a neighbouring rig] radioed: 'Where's the Ranger? We don't have her on radar.' We went to her location, where she was supposed to be, but there was nothing" (Kane, in Heffernan 2009: 69). "An icy feeling came over me then, one of dread" (Fahey, in Heffernan 2009: 74). "As we got closer, we started to come upon sparkles in the water. The realization then hit us that those were life jackets and men floating just below the surface—dead" (Fahey, in Heffernan 2009: 75). As one member of the Search and Rescue helicopter crew put it, "there was nothing moving out there except seagulls" (Sonntag, in Heffernan 2009: 81). The men on the neighbouring rig, the *Sedco 706* woke on the fifteenth to an empty sea where the Ranger had been. Max Ruelokke, part-owner of the dive company that employed Perry Morrison, articulates fears not prominently addressed in the inquiry report: that some men had hidden in an air tight part of the rig:

> There was a small void space next to the pump room. It was possible they could've survived there for some time.
>
> We tried signalling them by banging against the pontoons. No one banged back.
>
> Mobil asked that if we heard someone, if there was a noise, could we do something. We hastily contrived a plan which would've seen us burning a hole and getting a diver in there with an oxygen mast. At those depths and working within that timeline, it would've been extremely difficult. But not impossible.
>
> We gave up when it became apparent that even if someone had gone down there they would've succumbed to hypothermia. I never believed anyone would've done that; it would have taken a special kind of person. (in Heffernan 2009: 96)

Ruelokke insists that his men, the divers on the *Ocean Ranger* that night, would have worked with the rest of the crew, helping with the overall evacuation attempt instead of using their diving equipment and special skills to prolong their own lives.

Heffernan's interview with Ray Hawco, director of community relations for the Newfoundland and Labrador Petroleum Directorate, is the centre of *Rig's* political critique. Heffernan (2009: 113) presents Hawco's explanation:

> I met with the boys on a number of occasions, at a bar and a restaurant and someone's home. Sometimes there would be ten or twelve and other times more. There was no protection for them and they were thinking about starting up a union, but for the lack of something better

> hoped I'd take their message back to government and then on to the companies. The talked about the difficulties working offshore, but the worst message seemed to come from the *Ocean Ranger*: intimidation, a general lack of training on the drill floor, and inadequate safety drills. I told my superiors things weren't as they seemed.

Hawco was scheduled to fly out to the *Ocean Ranger* on February fourteenth, but he let another less lucky man have his seat in the overbooked helicopter.

With its more literary mode of inquiry, Heffernan's *Rig* allows its witnesses to point fingers:

> The place went dead silent when Premier Peckford sat down and started reading. Everyone expected him to appoint a Royal Commission to inquire into the loss of the rig. When he revealed the *Ranger* had listed just one week before, I thought I'd been punched in the stomach. The thing that sticks out more than the rest: "Conditions offshore are not severe enough to cause a shut-down." My God, I thought. Are lives worth nothing? (Phelan, in Heffernan 2009: 105)

Heffernan's inquiry also allows some of the "culprits" to speak for themselves. Susan Sherk describes how she became the public face for Mobil Oil during the *Ocean Ranger* disaster. She was hired by Mobil to do "people stuff" as the company found its feet: "They wanted someone who understood Newfoundland and could help translate the values of its people to their company" (Sherk, in Heffernan 2009: 108).[4] Like many family members, Sherk remembers vividly the announcement of the deaths: she smiled, by accident, for an instant. "If I could ever live my time over again, change that one thing I most regret, that would be it" (110).

Heffernan gives voice to perspectives that were isolated, muffled and repressed by "the" authoritative version of the inquiry report. More importantly, he validates a culture of memory-making through storytelling. As well, Heffernan offers a forum for people to tell their stories and to see that their individual nightmares, anxieties and rages were part of a collective trauma as well as very much their own. As the acting director of emergency measures at the time explains, he experienced the shock of seeing the face of his childhood friend when the drape slipped from a body as they transferred it from the dock to a hearse. He continues:

> Even today, if I'm tired or stressed out about something, I often wake from nightmares. It's the recurring dream of seeing that whole episode, his face, played out again in my mind like a movie. I never met his wife

> or his child, but it's stuck with me a lifetime. This is the first time I've spoken of it since then. (Browne, in Heffernan 2009: 118)

Heffernan exercises editorial control over the stories: the American oilmen disrespected Newfoundlanders, but more importantly, they disrespected the sea itself, and everybody had to pay. *Rig* is no wake, where, at least as the folklorists like to see it, people take turns and the living culture determines the order and duration of monologues. The "oral history" of *Rig* is closer to the creative non-fiction tradition of Studs Terkel than it is to the anthropological tradition of ethnography. Heffernan organized these recollections, and he did so in a politically potent way. *Rig* concludes with a claim to find closure in the collective identity of what I called earlier the "imagined community." Having heard the stories and seen the memorabilia people collected, Heffernan explains, "rather than sadness and defeat, I found in them courage, the need to overcome and find meaning in tragedy, that ineffable thing that makes us the unique people that we are, Newfoundlanders" (200). Folk history sentimentalizes the loss, but it also retains the potential to politicize it.

Lisa Moore's 2009 novel *February* also marks a turning point in the Newfoundland public's recovery from this collective trauma. *February* presents the *Ocean Ranger* disaster as the setting for exploring recovery from loss. At the same time, the novel expresses the spirit of the new Newfoundland, where recovery from the loss of the *Ocean Ranger* coincides with the birth of a new cosmopolitanism, though one still rooted in the community ethos of the old Newfoundland. The author is adamant about this: the novel is not "about" the *Ocean Ranger* disaster in any political, scholarly or even folk-historical way. In the widow Helen's gradual ability to imagine the final hours of her husband Cal's life in her own terms, the new Newfoundland is also able to recall the days before oil money from a safe vantage point. The promise of oil paid out, though the sacrifice was dear.

Moore's novel stabilizes the *Ocean Ranger* disaster as history, not in the sense of an objectively grounded retelling of what happened but in a sense that the culture itself is moving out of a more immediate traumatic past and towards long-term remembrance. This is not to say that the traumatic effects are eliminated; instead, their symbolic importance is modified in the very writing of this novel. Readers move with the protagonist as she works to differentiate her lost husband from her more existential sense of absence or lack.

The fictional widow reclaims her husband's story at the same time as she develops a new relationship with the man she will marry. Helen's personal working through emblemises the working through performed by the novel's readership, which includes the imagined "public of Newfoundland." Moore retells the well-worn story of the breaking window, the misinformed attempts to rectify the situation by the ill-trained men and Cal's final moments. The novel revises the technical chain of causation constructed by the 1984 Royal Commission of inquiry in the literary scholar Harold Bloom's sense of building on that account while at the same time claiming independence from it. It claims a direct psychic connection to the lost men that is somehow more primary and more affectively accurate than the technical story that the inquiry constructed from the traces the men left on the salvaged wreck of the rig. After a brief sketch of the report's version of the breaking portlight, Helen exhorts: "Imagine instead a man with his feet up" (Moore 2009: 149). We travel back to the human life of the rig with Helen, as she imagines the men's final meal, the men eating food like food from home, ignoring the fresh parsley sprigs. She "sees" her man playing cards. At this early point of recollection in the novel, the widow, like the inquiry report, now hits a block in the so-called chain of events: a wall of water washes the men from view.

Helen's unresolved mourning is a kind of fidelity. As she recalls: "That must be part of what they decided: If Cal died out there on the rig, Helen would never forget him. That was the promise. She will never forget him" (302). Her work in the story is to differentiate the promise to remember from a perverse promise to die with him, or at least, to not live without him. As the latency period ends and she knows that she is healing to the point where she can start a relationship with Barry—her conveniently sensitive and strapping carpenter—Helen thinks more and more clearly about a kind of division of labour. Cal dies and therefore he knows death; Helen must live and therefore she can know loss but not death itself.

At this early point, Moore leaves the evacuation to the next generation, to Cal and Helen's son, John. John works in the oil industry and he is university educated and well-travelled. His traumatic response to his father's death is to replay the terror of the water and to think directly—even brutally—about the so-called evacuation. The novel's narrative voice merges with the characters' attempts to enter the experience of the dying men, blurring clear subject positions. It is hard to know who the "I" would be all the way through this reflection:

> They'd found his father's glasses tucked away in his shirt pocket. John's father had taken off the glasses and put them in his pocket. He couldn't see a thing without his glasses. He must have stood on the deck as the rig was tipping, removed the glasses and put them in his shirt pocket, and then he probably jumped. His father would have had all his bones broken if he'd jumped from that height. But he might still have been alive when he hit, John thinks. John imagines he was alive. He has always imagined it that way. (139)

In *February* the neglect and callousness of the oil companies are comparable to the ferocity of the weather: this is the way things are. Moore's Newfoundland is still an enchanted world where the weather and human life are materially and symbolically linked. Bad weather is a material obstacle, and it is spirited, an agent expressing itself in cryptic ways.[5] The wave that breaks the portal is, for Helen, a "fist" that smashes into a quiet scene where a man is relaxing with his coffee after a nice meal.

Moore revises the romance of the grieving widow pacing the shore, in Helen's reflections on appropriate grief over a particular loss and its interplay with an existential sense of absence figured here as abandonment in the roiling deeps. The novel climaxes with Cal's widow finally able to imagine the evacuation. First, Helen can imagine that group of doomed men who incredibly launched a lifeboat and survived hurricane-force winds and waves only to capsize within crying distance of would-be rescuers. Then, more fundamentally, Helen is able to imagine the true end, the coming together of all the shattered parts, all the links in the "fatal chain of events" and "the obdurate wall of water" (301). Moore writes:

> She endeavours to face the true story.
>
> A crevasse forms in the cliff of water and it turns, as things sometimes turn, into concrete. Is it concrete or is it glass? It's mute and full of noise, angry and tranquil.
>
> How like itself and unlike anything else. (298)

Cal's death is radically unique and in a fundamental way, it belongs to him, and to him alone. The waves, however, and death in general, belong to everyone. Helen continues to face what she now presents to herself as the "true story": "This wave is death. When we say death we mean something we cannot say. The wave—because it is just water, after all, just water, just naked power, just force—the wave is a mirror image of death, not death itself; but it is advantageous not to glance that way. Avoid the mirror if you can. Cultivate an air of preoccupation. Get. Get out" (299). And, further, "The ocean is full of its own collapse, its destiny is to annihilate itself thoroughly, but for a brief moment it stands up straight. It assumes the pose of something that can last" (299). The water is a shapeless will, one that has been "working towards the chewing and swallowing of the world since the beginning of time," and a "great guzzling of itself is death, or whatever the end of life may be called," "it is unknowable" and cannot be named (300). Cal knows it. Helen does not, cannot and should no longer linger trying to know it vicariously through Cal. For Helen, working through the trauma of Cal's death and the loss of her children's father is a turn from past to future. She does not make this turn for the sake of others, not even for her children, but for its own sake and in its own time.

Even in the novel's titular month, February—the desolate month of snow, drear and loss—the hope of spring is present. The novel opens with Helen, out enjoying her grandson (John's baby, a surprise child), musing at the sparks flying upward from the skate-sharpener's stone, an image that echoes *The Book of Job*: "man is made to suffer as the sparks fly upward" (Job 5:7). Fragments of smashed light are scattered through Moore's descriptions. The novel moves back and forth through images of shattering glass: the portal of the ballast control room; the shattered mirror on Helen and Cal's wedding night; John, the boy child, smashing through a glass door. Helen's alienation is also figured in her distance from her children, as she puts on a show of engagement and yet observes their growing with detachment. As adults they are imposing and judging strangers. As one commentator puts it, Helen feels "not only robbed but banished" by Cal's death (Wylie 2010: 2).

Dreams express the characteristic fluidity of traumatic time. In Moore's *February*, dreams of the night of February 14–15 repeat themselves over the decades. One particularly unifying image comes from Cal's mother, who dreams repeatedly of a baby, exposed to the wind and weather, hanging outside in a tree. Her baby, Cal, risking himself to support his family, then dying at sea, is replaced in a metalepsis of recovery, by John's baby. In literary terms, a metalepsis is a trope that turns twice: first a particular is replaced with a general category and then that general category is, itself, turned into a new particular. In dying, Cal becomes one of the *Ocean Ranger*'s lost crewmen; his particular loss is taken over by the community's loss and recovery. It is not until a strange accidental child arrives that his family can reclaim Cal from the disaster. For Harold Bloom, metalepsis denotes the psychic defences of introjection and projection, that is to say, images that manage inner discord by representing it as taken in from outside (like a Trojan horse) or as a threat from the outside world (1975: 84). When used in inquiry reports and other bureaucratic writings, metalepsis is the administrative trope par excellence; it is the literary replacement that turns particular individuals first into a general category, and then turns that general category back on living people as a command to comply. Here, Moore's novel reclaims the moment of loss in a series of alternatives to the closure imposed on the disaster by the inquiry report. The lost man, Cal, is replaced (in a literary and psychic sense) by John's baby as well as by Barry, the new husband.

John and Jane's inconvenient baby transforms risk into a life-giving accident, outsiders into relatives and strangers into partners. Chance is transformed from a malignant technical risk that must be calculated and managed into a life-giving gift by Jane's unwanted pregnancy. Jane is an outsider, accidentally linked to their lives. Jane is an anthropologist before she becomes an accidental mother, and this saves Moore—the authorial self—from the discomfort of social science. As anyone who works with interviews knows, Victor Frankenstein is much more a social scientist than a natural scientist: we rummage around in people's graves, stealing parts of their life stories which we sew together in monstrous combinations. When we work with interviews we invariably break the poet W.H. Auden's law from "Under Which Lyre" (1946) that commands: "Thou shalt not sit/ with statisticians/ nor commit a social science." Moore protects herself from committing a social science, avoiding this first by not interviewing family members before writing the novel, though she did read the report carefully. Moore also reflects on the violence of appropriating the stories of others through Jane's ambivalence towards the project of capturing people's stories for academe. Moore's anthropologist-mother suspends the false creative life of collecting stories to embrace motherhood, choosing a baby over the production of a scholarly report.

February concludes with a healing story, as the administrative replacements of the inquiry report are themselves replaced with literary revisions. The Royal Commission report turns the desperate men in that ballast control room that night into "untrained workers" who were hired under a potentially problematic local hiring policy that was intended to secure Newfoundlanders their share of offshore jobs. The full tragic potential of the *Ocean Ranger* story lies in this: that the very aspiration to break out of the colonial past by asserting Newfoundlanders' rights to jobs in the offshore without demanding training programs and safety protocols actually repeated the pattern of exploitation that the province's politicians and company bosses had been so anxious to avoid. The report then turns those "untrained workers" back into new policy imperatives for health and safety training, where the "untrained worker" becomes a source of risk in the workplace and must be disciplined into a more thorough self-consciousness about his work and even his deportment. The novel picks up the report's version of the chain of events but revises it even further by reclaiming that untrained worker back, turning the generic "*Ocean Ranger* victim" back into a particular lost character. This character, Cal, dies an individual death, is mourned by another particular character, and is then turned into a happy memory as his widow transforms herself into a new wife. The dead man is replaced by, or rather supplemented with a living man, while a memory is supplemented with a future.

It is this capacity to bring readers back into a direct emotional relation with the lost man and his survivors that makes literature one of the central genres of memory work. A novel allows us to spend time with the ghosts that haunt us in a way that problematizes the boundary between then and now. Literature enables us to visit with the lost men and with the "us" of an earlier time without being trapped in the vicarious death of reliving the past. Literature also dispels the illusion of objectivist narratives that claim that we can be dispassionate writers and readers of history and that the people we have lost and the events of the past are simply objects out there, waiting for us to collect them.

When Helen's new husband goes swimming in the closing pages of *February*, readers are invited to revisit our own losses. Barry disappears into the sea, and we join Helen in scanning the horizon anxiously. When we find him again, this second husband, climbing up the beach to greet Helen, we are invited to leave Cal to his private death without forgetting him. *February* is thus an expression of a public working-through, and as such it also participates in a potentially dangerous form of forgetting. The danger of the literary here is that it does privatize or personalize the loss and so it stands to diffuse residual political anger. The novel *February* marks the end of a period of latency in the aftermath of the collective trauma of the *Ocean Ranger* disaster. This means that the imagined community, the character of "the people of Newfoundland," is able to revisit the loss without repeating the shock.

In *February*, then, Lisa Moore gives voice to the public working-through of the *Ocean Ranger* disaster, that is, its further revision from a "learning story" into a symbol of the new Newfoundland, the Newfoundland of a proud and independent people, a people that has wrested material benefits and recognition from the most powerful corporations in the world. Neither Helen, nor anyone in her family, focuses on the pursuit of justice, let alone on vengeance. Life for them is bound up with the oil industry and the practicalities of money. Moore's novel deals very little with explicit questions of justice or legal practicalities. While Helen knows the inquiry's "findings" about the "chain" of events that caused the disaster, she does not participate in the inquiry hearings and does not register the unfolding of that testimony at all. As Helen is able to remarry, so Newfoundland has reforged relationships with the oil industry and turned towards a new future. The broken promise of the early 1980s, exemplified in the Cadillac of rigs, the *Ranger*, could not simply be reforged. A new promise was needed, this one showing the new relations between industry, government and the public. The *Ocean Ranger* persists, nonetheless, as a haunting symbol of the bad old days and of the constant anxiety that the present calm conceals an apocalyptic potential. As Helen's new man remarks while they watch the New Year's celebrations over St John's bustling harbour: "Imagine if a spark from them fireworks landed on those oil tanks" (383).

The central critical power of the novel and the oral histories is also their healing impulse, which is to say, the revision of the story of what happened, who was to blame and what was to be done about it that the Royal Commission so authoritatively told almost three decades ago. The 2009 versions tell the story in terms of the living community, and so they give voice to a healing, as well as to an enduring anxiety about the capacity of Newfoundland culture to manage the new promise of oil. Neither book fully engages the anger that some family members sustain as they try to understand what happened to them when they were thrown into a struggle for accountability with some of the most powerful organizations in the world. Moore's novel does not deal with the lawsuits, except to mention them in passing. Helen is so deeply immersed in her loss and the traumatic distance between her inner life and the day-to-day work of raising her four children that the settlement process becomes what it must have been for so many young widows: a strictly practical affair handled by others. Heffernan recounts some of the hubris of central managers, particularly Jimmy Counts, but leaves the systemic problem of how it came to pass that they were operating without adequate regulation unexplored.

THE POLITICS OF COMMEMORATION

Commemorations are practical activities by which people and collectivities transform the inner turmoil of loss into outward expressions of remembrance. Commemorations are "attempts at closure, at decisiveness and imposition, like the sharp report of a field gun at a military commemoration and the ringing sound that follows it: this is the sound of remembrance, this the silence" (Sider and Smith 1997: 7). Punctuating the time of public trauma, commemorations close off some lines of thought and dialogue so that fresh ones can spring up. Debates about public commemorations and the interplays between them reveal the local tensions that invigorate a community. As we have seen, the inquiry report makes sense of the traumatic event so that the way forward is to forget the regulatory failure and political betrayal at the disaster's root and to immerse in the story of technological and economic progress. The other forms of commemoration I have just described challenge the bureaucratic orderliness that this official story aims to impose (Sider and Smith 1997: 11, 13).

Practices of commemoration interact with formal legal and historical writings in a network of remembrance, as we will see in the following chapters. Static documents anchored in "facts" legitimated by liberal legal processes and peer review are reused, resisted and revised in popular versions of the Ocean Ranger story. Legal documents and commemorative practices interact in the memories of individual people and "the people of Newfoundland," a symbolic character that is imaginary in one sense and foundational for many individuals in another (Anderson 2003). The ways people describe and commemorate the Ocean Ranger disaster have changed over the three decades since the loss. The initial version, where the betrayal of trust was recollected as a necessary lesson, harmonized with political story lines already in play, particularly with

Newfoundland nationalism and Canadian paternalism towards the "have-not province." The imagined communities of Newfoundland and Canada were injured by the disaster and then reconfigured during the revisionary processes of the aftermath. First, the Ocean Ranger disaster shifted from a story of government neglect and corporate predation into the technological learning story of the report of the Royal Commission of inquiry. Then the technological learning story of the report was itself revised decades later when Newfoundland writers Moore and Heffernan recast the Ocean Ranger as an emblem of the old Newfoundland. As the story goes, the Newfoundland of the *Ocean Ranger* disaster is gone. In its place stands a new Newfoundland, one that triumphed over its colonial past by standing up to Ottawa and to oil companies so as to seize control of their destiny by realizing the promise of oil.

NOTES

1. Mobil collected these telexes into a bound book, with "In Memoriam" and Mobil's logo on the cover. The inside cover is a line drawing of the rig and "In Memoriam, February 15, 1982" with Mobil's logo repeated. The book opens with the men's names and continues with telexes of condolences from Queen Elizabeth, Queen Beatrice, ODECO, Schlumberger and many more. It is a strangely nice little book, sent to all families by Mobil Canada.
2. As E.J. Pratt puts it:

   It took the sea a thousand years
   A thousand years to trace
   The granite features of this cliff
   In crag and scarp and base.

   It took the sea an hour one night
   An hour of storm to place
   The granite features of this cliff
   Upon a woman's face.
   (in Earle 1998: 99)

3. "When he came home after that first trip, he had a copy of the telex in his pocket. First of all he said the toolpusher and the captain came to the radio room while he happened to be on duty at the time, and both of them stood there arguing over this particular telex. The captain was in favour of sending it and the toolpusher was not. What it was about was that some time previous to that they had had an oil spill which I don't think the media or the families had been informed about. The captain thought it should be made know, that it should have been reported to the Mobil/ODECO people here on shore. And Greg said that they argued about it back and forth and so in the end he was not allowed to send it" (Hickey 1984).
4. "Mobil Corp. has brought to the East Coast its experience in operating in over

100 countries. One indication of its shrewdness has been in the public relations and social impact field. This has been particularly challenging in Newfoundland, with its recent history of suspicion of multinational corporations. Minor conflicts have occurred between Mobil and various community groups, but, for the most part, they have been managed with minimal fallout for the company. In large measure this has been due to the work of Susan Sherk, Mobil's public affairs chief. The company was astute in hiring her for this delicate position. A graduate of a prestigious New England college, Sherk spent two years doing anthropological work among Larador native peoples. This was followed by a five-year stint as editor of Memorial University Extension Service's DECKS AWASH, a monthly journal produced in a popular format dealing with rural issues in Newfoundland. Sherk has presided over innumerable social impact consultations with every variety of community and professional organization in Newfoundland.... At one such gathering in St. John's in 1981 she claimed that, whereas in the past Mobil might have been primarily motivated by profit, the corporation was in Newfoundland to develop Hibernia for the benefit of Newfoundlanders as much as for its own financial well-being. In 1987 Sherk was promoted to Mobil's head office in New York" (O'Neill 1988: 157).

5. This revises a longstanding sense of the ocean as a difficult, life-giving and life-taking companion, captured here by Farley Mowat in *This Rock Within the Sea:* "Perhaps no other ocean emanates such a disturbing feeling of sentience. It is not just a realm of water, it is a presence —one of incalculable moods. It is seldom still. Even in its rare moments of brooding calm, a long and rhythmic swell rolls under the surface so that it ripples like the hide of a monster. It is at times such as these that men of the sea distrust it most... 'weather breeders' they call such days, and they prepare for what they know will follow: a passionless and almost inconceivable violence of wind and weather" (in Earle 1998: 102).

5

REMAKING THE PROMISE OF OIL: THE AFTERMATH OF THE OCEAN RANGER DISASTER

"If you want a happy ending, you need to know when to end your story."
—Premier Danny Williams, on his retirement from public office, quoting the author, actor and filmmaker, Orson Wells

Several questions emerged out of my long-standing reflections on the *Ocean Ranger* loss. These questions link the regulatory failure at the root of the disaster to liberal capitalism's capacity to encourage us to forget legitimation crises. The forgetfulness of liberal capitalism is expressed in, and sometimes resisted by, the ways we work through personal and collective traumas. My guiding questions have been: How did those companies come to operate in Canadian waters, with a Canadian crew, without any effective government regulation? How was it that the only punishment the corporations faced were the modest settlements they paid to avoid lawsuits launched by grieving and in some cases destitute families? How has the *Ocean Ranger* story become what it is for many people: a tragic narrative of wind and waves, of technological ambition and economic pioneering? How did the social and political processes of the aftermath help to secure public confidence in the promise that offshore oil development was the key to Newfoundland's triumphing over its colonial past? What role, if any, did the "sacrifice" of these men play in subsequent determinations to wrest recognition and benefits from transnational corporations on behalf of the people of Newfoundland and Canada? What is the relation between "my" disaster and other disasters, especially disasters caused by regulatory failures, like the *Deepwater Horizon* explosion—the Gulf oil spill—and even the mortgage crisis faced by millions of Americans? That I persist in asking these questions all these decades after the *Ocean Ranger* disaster is a sure sign that there is something wrong with me. Staying this kind of crazy is hard work—those of us who do it pit ourselves against a wide range of social, political and cultural processes that manage potential legitimation crises whenever they threaten.

When an unforeseen and initially inexplicable event like the loss of the eighty-four men and the state-of-the-art *Ocean Ranger* breaks in on normal life, we try, as a community, to make sense of what has happened. We look back and tell ourselves that this is a limited term: we will pause for a while to investigate the past, and then we will return to projects aimed at the future. To settle on an account of what happened and who is to blame, is to make the turn to what is to be done; to tell the story and assign responsibility is to step out of forensic mode into free creativity. But this tidy model of shock, forensic work and then a clearly delineated turn to the future is a fantasy. This is not how it works, especially when the shock rattles the bonds of community itself. In this chapter I map out the processes I described in earlier chapters, comparing their relative claims to objectivity and to evoking feeling. The process of a formal inquiry, like the Royal Commission, produces a report. That report makes a strong claim to representing reality objectively; it is what the historiographer and psychoanalyst Dominick LaCapra calls an "objectivist" or "documentary" account. Such a report has to convince readers that it literally provides a window onto an empirically accessible past (LaCapra 2001). In contrast, a novel like Moore's *February* claims to express a different kind of reality. It is fiction—Moore is adamant about this—and it provides a window onto a more emotional kind of reality. It is not held to account for factual accuracy in the same way as the inquiry report, though it must meet intuitive expectations of plausibility. Both *Rig* and *February* are more "constructivist" accounts than the inquiry report is. This means that they are up front about the "factors that 'construct' structures—stories, plots, arguments, interpretations, explanations—in which referential statements are embedded and take on meaning and significance" (LaCapra 2001: 1).

Aftermath texts move between these two poles—an objectivist, positivist, "documentary" approach, which we can see as a naive empiricism that understands itself as gathering facts and presenting them objectively to a rational audience, and a "radical constructivism," which draws attention to the ways that referential writing is framed and to the figures of speech every text uses to draw the reader in. Each of the processes of the *Ocean Ranger* aftermath that I look at here describes and interprets the disaster by holding these two sides—the event and the structure of its recollection—together in a way that presents the disaster as a kind of story that readers can understand. At the same time, texts take up and revise images from each other and so they are bound together and interrelated.

The terms we use thirty years after the event are not those we used in the first day. In fact, the aftermath is a "revisionary process"; the working of emotions into expressible forms is a process of replacement (Bloom 1975; Green 1982; Dodd 2001). A feeling or affect is replaced by a representation that cannot communicate it literally, but can evoke it, or at least evoke parallel feelings in other people. Tracking such "revisions" is the key to mapping the ways the legitimation crisis is managed in the aftermath of an industrial disaster. For example, the official Day of Mourning of February 19th that commemorated the loss of the eighty-four men was repeated in a revised form three months later, with the Day of Mourning that marked the referral of the jurisdiction question to the Supreme Court of Canada, which was presented as the federal government's exertion of power over Newfoundland's sovereignty.

In previous chapters I touched on seven main socio-political processes that respond to a potential legitimation crisis and in which we work through the collective trauma more or less authentically. In each of these, we produce works, ways of representing our loss and our pathway into the future. These are:

1. quasi-judicial inquiries and their reports;
2. mass media and the power to buy "the news";
3. electoral politics and ads, leaflets, platforms, speeches and election results;
4. tort lawsuits and their settlements;
5. parliamentary debate and legislation;
6. commemorations and monuments; and
7. popular art, performances and literary works.

Each makes some claim to representing reality, though only the first five in my list make any strong claim to objectivity. If there had been criminal charges and a trial, that process would have made the strongest objective claim: it has to because the "work" produced by the criminal process is the verdict. And the verdict justifies the modern liberal state's monopoly on enacting violence in the wake of wrongdoing. "Objectivity requires checks and resistances to full identification, and this is one important function of meticulous research, contextualisation, and the attempt to be as attentive as possible to the voices of others whose alterity is recognised" (LaCapra 2001: 40). Because such a narrative claims to be a transparent window onto objective reality, the traumatic symptoms are repressed.

Each aftermath process and the kind of work we produce from it revises the experience of the disaster into a form that we can grasp without being thrown back into the original shock of the loss. The extreme emotions evoked by sudden loss complicate public access to facts and so also to pathways for reform. The literary figures that turn individuals into general categories and back again are the work of recovery and of repression. One measure of such texts is the kind of narrative claim the text makes.

QUASI-JUDICIAL INQUIRIES AND THEIR REPORTS

The socio-political power of the Royal Commission on the *Ocean Ranger* Marine Disaster can hardly be overstated. From the time the inquiry was announced until the release of the report, everyone was asked to suspend judgment: all the facts had to be aired and all the players deserved their chance to set the record straight. This produced a cooling-off period in the earliest moments of the aftermath. More importantly, it presented the spectacle of the state holding corporate agents accountable to "the public." From the time the report was released, nobody can write credibly about the disaster without referring to it. As important as the recommendations for reform are, the report's real power lies in framing the historical memory of the disaster. Threads from the inquiry report's version of the event are woven into every other aftermath account. The inquiry and its reports literally closed off critical lines of questioning for three decades following the disaster. Newfoundland took care of the *Ocean Ranger* aftermath in an overtly paternalistic sense. Not only did the House of Assembly use public funds to support family members so that the community could maximize the payout from the oil companies, but the entire community was implicated, and constructed, in the fact-finding, testimony and even in the final writing of the *Ocean Ranger* report. "Report One: The Loss of the Semisubmersible Drill Rig *Ocean Ranger* and Its Crew" masquerades as happy naive empiricism. No one survived the *Ocean Ranger* disaster. There are no witnesses. The report steers away from this problem, just as it does from the engagement of readers' empathy with the doomed men.

The Commission of Inquiry is a hybrid of evidence-based fact-finding and more open witnessing forums like truth and reconciliation commissions. The testimony of quasi-judicial inquiries brings those responsible for the disaster face-to-face with the people's representatives. Inquiry hearings dramatize the state's ability to produce truth from witnesses. Witnesses are freer to speculate and are less constrained by strict rules of evidence. Such testimony is at once more open and less accountable than testimony in a more formal judicial process. Inquiry reports address a looming crisis in state legitimacy more or less directly. At the same time, they tend to impose a premature closure on the forensic attitude of the aftermath. They punctuate a turn from reviewing the past to acting in and for the future.

The inquiry report draws on a convention of social-scientific policy reports reaching back to the poor law reports of the nineteenth century, as well as on language developed in mass media "news." It is related to, but not determined by, the self-consciously ideological work of public relations campaigns. It addresses a legitimation problem that those campaigns do not. That is to say, the inquiry report addresses the persisting sense that "the public," in this case

"Newfoundland," deserves recognition but also that it can be constructed in the course of retelling the traumatic story. These processes and their interrelations are deeply ambivalent. Of course, to the extent that governments and corporations simply buy the airwaves, public memory is shaped by a simple and naked exercise of force. I find it encouraging, though, that more complicated processes of recognition and repression are also needed to restore at least the fantasy of a united national agent.

In the testimony, the fact-finding and recommendations of Volume One of the report of the Royal Commission and in a conference on safety in the offshore, the Royal Commission of inquiry effects its own series of revisions. The conflict between local interests and the interests of the international petroleum industry that lay at the root of the *Ocean Ranger* disaster is presented as a confusion over regulatory jurisdiction between Newfoundland and the federal government. Government complicity with corporate negligence was recast as a shared lack of technical know-how. The report uses an extended version of the metaphor "loss is a gift": we see that "disaster is education," "disaster is evolution" and even "disaster is progress." The oil companies' shoddy operation was eclipsed by confidence in Newfoundland's, Canada's and the oil industry's ability to deploy science and law to discover the truth, to report on it and to set a course for the future. Calls for punishment became recommendations for reform.

The report creates the character of the Commission itself, reifying the testimony and cross-examination process into a single-minded agent of reform. The Commission is a scientist of sorts, one who marshals a vast collection of different kinds of facts brought together by the procedures of validating evidence. The dominant images of the report are of volume being distilled, retrieving truths from depths, marrying data and talk, and bringing individuals and collectives into constructive relations. A text is its own little world, one that demands to be understood on the terms it presents to readers. The reader has to be able to recognize the kind of "reality" a text constructs in order to play her role as the one who brings the words to life, in a sense. The "reality" that an inquiry report constitutes is:

- a present context—in the *Ocean Ranger* case this is presented as an industry-wide lack of knowledge about regulation and how to drill in extreme conditions and not a failure of regulation. In the report, our governments' failure to regulate and the American corporations' refusal to self-regulate are retold as a knowledge problem;
- a possible-projected context—where industry and government should collaborate on technological discovery so that oil production can continue

to evolve to survive in its ever-changing environment;

- a plausible agent of intervention—the Commission itself will reform government regulators, industry self-regulators and imprudent workers;
- a causal narrative or syntax to situate the actors in the report's present context, poised for future action—the report publishes recommendations and a time line for their realization;
- a reader-observer to perform the synthesis necessary to syntax—"the public" that make up Justice Hickman's target audience, that is, as I said earlier, a Newfoundland in Canada and a Canada in the international oil industry.

It is interesting that some of the "evidence" that the Commission used to reconstruct the last hours of the men's lives on the rig was sent to the Canada Science and Technology Museum in Ottawa by the Commission secretary, David Grenville. This move exemplifies the connected nature of all these ways of remembering: Grenville intended to "protect the integrity" of the evidence in case it was needed during the families' litigation, and he also believed that the collection had "historic and technological significance" (Grenville, in Babaian 2005: 69). The evidence that Justice Hickman contentiously salvaged "for the public" should now be kept in case it was needed in the private suits but also for the future. The items are the same, but their meaning is transformed in their transferral from the quasi-judicial inquiry to the museum. Now, they represent a perplexing possibility: should there be a show of the artefacts? Would anyone be interested? How could these cryptic fragments of the giant wreck lying now in "a designated dumping spot" be made legible and interesting for viewers? Given the sensational story of loss, maybe this evidence could, with proper explanation, have "some numinous power"? (Babaian 2005: 72). Such an exhibition could go any number of ways: it could be framed to reinforce the dominant story of a "lesson learned," or it could be turned into a drama about the perilous ocean on Valentine's Day, in the spirit of some of the post-movie *Titanic* museum exhibitions. Most interesting might be an exhibit showing both the techno-organizational failure at the beginning of the narrowly defined chain of events and the forensic process of determining what happened. It might be interesting to see the broken portal and ballast control panel and also artefacts from the Commission itself, notably the scale model of the rig along with commentaries about the reconstruction of an event out of mute technical relics.

MASS MEDIA AND THE POWER TO BUY "THE NEWS"

Public relations campaigns are the blunt instruments of disaster's aftermath. It is by now a cliché to note the extent to which corporations can buy the airwaves. The following sketch illustrates some of the most obvious moves a company makes in the immediate aftermath of an industrial disaster with its name on it.

Start strategically:

- immediately focus on technical glitches as the "cause" of the disaster;
- emphasize the radical unlikelihood of the fatal sequence of events, with the trigger point identified as a particular worker or group of workers (preferably dead ones who cannot testify to the pressures and expectations of their managers); and
- claim incessantly that the companies were working within the regulatory strictures as they understood them and thereby implicate all levels of government in the disaster.

Problematize evidence:

- muddy the scientific waters with multiple "expert" accounts; and
- minimize the financial impact on the corporate person by rushing financial settlements so the claimants do not have the benefit of forensic studies.

Restructure:

- divest of consumer brands to avoid boycotts;
- pay shareholders dividends before the lawsuits strike;
- scapegoat local workers; and
- scapegoat a prominent manager.

Replace local experience with transnational myth:

- take advantage of victims' need for closure, the public's waning attention and shareholders' remarkable faith in the magic of investment returns;
- exploit any prejudices against the local people;
- counter public reports with industry-produced science; and
- emphasize the essential role of free enterprise in leaving the "good corporate citizen" to flexibly self-regulate according to its exclusive expertise.

The encouraging news from trauma studies is that even the most sophisticated public opinion research cannot be certain to produce a successful public relations campaign. Trauma means contingency. The traumatic impact of a disaster can both facilitate and obstruct the self-conscious deployment of ideological strategies. Some traumatized people resist closure, refusing the comforts offered by redemptive narratives of "everything happens for a reason" or "every disaster is a lesson learned." My need to write this book, Scotty Morrison's feeling sick thinking about their settlement and all the traumatized people in the "madhouse" of Mike Heffernan's *Rig* share a resistance to such redemptive narratives. This refusal of closure can be politically effective, particularly for underdogs. A rational sense of injustice fuelled by obsessive focus and a refusal of comforting stories may be *the* definition of a critical attitude.

While I did archival research into media coverage of the early days of the *Ocean Ranger*'s aftermath and the 1982 Newfoundland provincial campaign, I was fortunately able to borrow Marie Wadden's files. As the CBC television reporter who covered the inquiry, Wadden and her files are a wealth of information. Not only am I predisposed to appreciate her coverage because of this generosity, I also cannot help but contrast it with the buffoonery and techno-dazzle of the coverage of the attempts to cap the Macondo well in 2010. There is much more work to be done to analyze the impoverishment of reporting in the past thirty years.

Of course there will be no regulatory reform unless there is public awareness of the need for it, and so news reporting is key. Along with reporters from the *Evening Telegram,* Wadden guided Newfoundlanders and Canadians through the inquiries that followed the disaster. These reporters played a crucial role in providing "the public" with a sense of involvement in the inquiry and so also in virtually having a hand in writing the report, as I argue in Chapter One.

ELECTORAL POLITICS – ADS, LEAFLETS, PLATFORMS, SPEECHES AND ELECTION RESULTS

I began this chapter with an excerpt from Danny Williams' retirement speech. By quoting Orson Wells, Williams applauds himself for recognizing the theatrical nature of his political success. Of all the ways of managing potential legitimation crises that we consider here, the rhetoric of people as they seek office is the most self-consciously constructivist. Politicians use platforms to build on voters' existing positive impressions of them. In this sense, platforms and campaign policy statements both are and are not about their content. The content reinforces the character development. The rhetoric of political campaigning suppresses facts and "switches the channel" to divert affective energies. When party strategists sullenly complain that they are misunderstood by the electorate and gesture at "the facts," the real story is that they have lost their sense that they can direct public discourse. In 1982, Premier Peckford constructed his image as a fighter for "the people of Newfoundland," as a symbol of self-governance and a new kind of maturity on the Canadian and international stages. The anger and anxiety generated by the traumatic loss of the *Ocean Ranger* was diverted through the rhetoric of job creation and

Newfoundland self-determination. It is worth repeating that "Ottawa," rather than the American oil companies, was cast as the culprit of the tragedy.[1] The Newfoundland cultural identity was reinforced by the crisis of the *Ocean Ranger* loss; the "old" Newfoundland had to give way to the new, and Brian Peckford personified a determination to prevent past subjugation to wind, waves and colonial power from swallowing the future.

The province of Newfoundland did not start reporting on elections financing until 1993, so there are no records of who bankrolled the 1982 campaign.

THE THREAT OF LAWSUITS AND FINANCIAL SETTLEMENTS

For families, legal negotiation towards financial settlements is a paradox; it is a radically alienating struggle with a partially liberating outcome. Few people understand what happens to their "case," and yet they can become so embroiled in the mysterious, quasi-logical parry and thrust of negotiations that settlement is a relief. Blood money relieves family members of the responsibility to avenge the death, and it decreases the pressure when the "accused" cannot be held accountable through criminal charges, either because of complex organizational networks or because of a failure of political will. As I argued in Chapter Three, on the history of blood money, the trade-off of the modern age was to give up the right to vengeance, to have the punitive side of the blood-feud managed by the sovereign. If the sovereign no longer holds up its end in punishing massive corporations for their wrongdoing against individuals, communities and the environment, what then? This is a naive formulation of a complex matter, I know, but I still find it hard to believe that it was impossible to prove criminal negligence in the *Ocean Ranger* case: were the laws and political will simply not there? Or was the promise of oil so fragile that authorities backed off, leaving the only punishment to be meted out by grieving women and men who rolled the dice with international tort lawyers?

The translation of unique irreplaceable loss into a cash payment gives shape to unrepresentable suffering and so offers an element of relief from the indeterminacy of loss. At the same time, the stigma of accepting cash for life transforms surviving victims' social positions. Family members lose their status as inexplicably chosen for suffering and exiled by injustice. They are forcefully reintegrated into the exchanges of ordinary life, and, like medieval usurers, they receive money without work. The acceptance of a financial settlement speeds the rate at which the community becomes impatient with survivors' demands. The exchange of blood money is a moment of closure for the community that is purchased at the cost of imposing false closure on family members. The oil companies were shamed rather than punished, and it fell to grieving family members, especially to women, to struggle to satisfy the demands of justice.

Receipt of money, and especially the settlements from threatened lawsuits, radically narrowed friends' and families' quest for justice.

The ambivalently crushing and liberating moment when families accept a financial settlement from the companies is one of the most extreme impositions of what LaCapra calls an objectivist account. The radical uniqueness of the lost man, the dignity of his humanity in a sense, is equated with a cash payment. This is a distillation of a narrative of loss and replacement into market equivalence. Having capitulated to rendering the lost man in dollar terms, family members step gradually back from public debate. Their role as victims is transformed, privatized.

The international personal injuries lawyer Benton Musselwhite personifies the ambiguity of individual citizens having the right to sue massive corporations for "damages." In his recent work with Dole banana workers in South and Central America, Musselwhite was once again accused of unethical soliciting and spreading misinformation. Dropping these charges is reportedly one of the terms of the out-of-court financial settlement between Dole and the workers (*Texas Lawyer* 2011).[2] Musselwhite consistently casts such charges as smear campaigns by big companies and their supporters. Musselwhite's charm and zeal for holding corporations accountable also happen to be exceptionally lucrative. Worse things could happen to a grieving family than to fall in with such an ambitious, experienced advocate when they are staring down the barrel of a fully loaded corporate legal strategy. And, considering that the dollars wrested from big companies in tort negotiations are still the only form of punishment they are likely to face for wrongful death and environmental degradation, why not? Why not brave the charges of trying to make money "off the backs" of the dead? Why not take it to the point of the family's psychological limit?

Talking to the American lawyers I interviewed made me feel very Canadian. That was particularly true during my conversation with Musselwhite. Civil lawsuits seem to me to be necessary evils. It is what we settle for as we work to improve the means of holding corporate agents responsible collectively through criminal law and fines for breaches of robust safety and environmental regulations. Tort is inadequate as a form of punishment against huge companies. More than this, it places an unconscionable burden on grieving family members to pursue civil suits as far as they can precisely because there is no collective sanction.

When Benton Musselwhite told me that our $25,000 settlement was "a dishonour to your brother's memory," I think he misunderstood something about Canadians' understandings of the purposes of such lawsuits. We regulate, Americans sue. Canadians presuppose that government is responsible for punishing individuals and corporations, be that through jail, fines or even holding them accountable in public hearings. We do not tend to begin from the presupposition that our right to sue is also a responsibility to the rest of the community. For Canadians, civil suits are private in a full sense. Americans are different. They may in fact see their lawsuits as an ethical obligation. A major problem with contemporary tort law is that it pits a particular human being, a biological and psychic individual, against a fictional being—the corporate person, which is a sub-culture in its own right.

If insurance money represented a family's prudence in buying into a statistical evaluation of risk, Workers' Compensation represented an industry's and a community's prudence and a significant, if limited, means of encouraging self-regulation. What, then, does money from civil lawsuits represent? This is a live question in Canadian legal culture. In the past couple of decades, Canadian settlements have started to provide damages for harms done beyond economic loss, most notably in cases where the injury emerges from systemic violence with deep psychological effects, as in Donald Marshall's wrongful conviction and incarceration in Nova Scotia, or the abuse of boys in the "care" of the Christian Brothers of Mount Cashel. Some personal injuries lawyers are lobbying to expand Canadian tort practices so that courts take more proactive measures by awarding damages intended to modify corporate behaviour. This strikes me as inadequate: what calculus could determine the sum ODECO, Mobil and Schlumberger should pay to make them more mindful of their workers' lives? Whatever the sum, should this not be paid to the state, and not directly to grieving families?

PARLIAMENTARY DEBATE AND LEGISLATION

Jim and his co-workers died because our governments failed to take reasonable precautions to protect them from the profit-seeking of the oil industry. This is the real root cause of the *Ocean Ranger* disaster. To expect corporations to self-regulate in any way that compromises the most direct path to profit is naive. A major outcome of the *Ocean Ranger* inquiry was the formation of a joint regulatory agency with the signing of the Atlantic Accord, the Canada–Newfoundland and Labrador Offshore Petroleum Board (CNLOPB), a regulatory regime, as well as a spin-off industry in safety training. That said, the Newfoundland offshore is notoriously secretive (Hart 2005) and more inaccessible to journalists now than ever; as CBC's Marie Wadden told me: "I tried to get to talk to the president of ExxonMobil but I was literally body-blocked by his people" (2010). Furthermore, as became evident in the Wells inquiry into the loss of the Cougar Flight 491, the CNLOPB still combines oil industry development with safety and environmental regulation in a relationship that even Wells considered to be too cozy for comfort.

Corporations are responsible to their shareholders, so they turn the opportunities afforded by their environments into profits for people who often do not live in, or near, the areas most effected by the risks taken by industrial development. The shareholder's risk is strictly a financial one as they are protected by a "corporate veil" of limited liability. Government's job is to define those opportunities so that corporate activity also benefits the host community through job creation and spin-offs. At the very minimum, governments are responsible for securing our lives from any reasonably foreseeable threat.

John Crosbie, MP for St John's West and minister of numerous federal portfolios, championed the recommendations of the *Ocean Ranger* inquiry, as far as Justice Hickman is concerned. One recommendation that remained unfulfilled was, as I have mentioned, the commitment of government and industry to collaborate on serious research and development around the evacuation of crew members from the rigs. Another recommendation, still contested, was the stationing of a rescue helicopter and crew in St John's, instead of relying on Search and Rescue from Gander and Greenwood (Bailey 2011). Further, as the lawyer for unionized workers at the Hibernia and Terra Nova oilfields testified at the 2010 inquiry into the preventable and fatal crash of Cougar Flight 491: CNLOPB regulators have "displayed institutional lethargy that's absolutely mind-boggling" (*Telegram* 2010).

To the best of my knowledge, extending criminal charges to corporate negligence was not seriously considered in the wake of the *Ocean Ranger* loss, though some parliamentary rhetoric was extremely heated and accusatory. Parliament passed the "Westray Bill," Bill C-45, in March 31, 2004, and this allows negligence to be attributed to the corporation as an organization as a whole so that intention and deeds do not have to be traced to a single human individual. The major gain of the Westray Bill was to make occupational health and safety negligence a criminal rather than a regulatory and civil offence.[3] Keeping the dominant "learning story" of the *Ocean Ranger* aftermath in mind, it is worth noting that the attempt to criminalize corporate negligence did not bear legislative fruit until the public outcry against the 1991 Westray explosion and deaths, supplemented by the United Steelworkers of America, became a lobbying force in the constituency of a powerful member of the Conservative government. Whether senior executives will be held accountable under this law remains to be seen.

COMMEMORATIONS AND MONUMENTS

Mourning people give shape to their loss in sculptures and other means of transforming the physical world into symbolic spaces. Memorial services and physical monuments often commemorate loss in ways that reinforce dominant power relations. However, they can also become sites of resistance where people challenge monolithic, redemptive stories like that presented by the inquiry report, where the dead are portrayed as sacrifices to a noble collective good (Edkins 2003; Kaplan 2005).

The *Ocean Ranger* monument that was crafted by the Ferryland sculptor A. Stewart Montgomerie is ambiguous; it evokes rather than represents an anchor. At the same time, it gestures towards the February ocean, suggesting an iceburg. I was surprised by my strong response to both the memorial sculpture and the photograph that Montgomerie submitted to the competition to choose the artist for the monument. As an anchor, the sculpture evokes solidity, and yet it is out of place, solitary and exposed. Montgomerie's *Anchor* captures the desolate sacrifice of the *Ocean Ranger*'s crew in the name of progress, and yet its vulnerable isolation suggests the question: Was it worth it?

The tablet listing the men's names stabilizes the ambiguity of the sculpture in the particularity of the men. There is a strange power in a list of names. When we know one of the people on the list, that name links us emotionally to the fullness of a life and the enormity of that particular loss. This invests the names of the people we did not know with mysterious volume: we know them only as gone.

POPULAR ART, PERFORMANCES AND LITERARY WORKS

Having dealt with these at length in the previous chapter, I will simply sharpen the point that collectivities can undergo periods of latency in the wake of a trauma. In 2009, two major works represented and revised the *Ocean Ranger* disaster from the perspectives of ordinary life. Why, I wondered, had it taken almost thirty years before Newfoundland writers tackled full-length accounts of the disaster from the perspective of those who suffered it? Fiction and oral history are both open to traumatic memory precisely because their authority rests in their affect, in feeling, and not in any objective claims to accurately refer to empirical facts. The publications of *Rig* and *February* marked an end to at least a phase of latency, "the period during which the effects of the experience are not apparent" (Caruth 1995: 7). As well, these texts offered alternatives to the shock-forensics-action narrative of the inquiry report, and so they challenged "the theoretically regulated order" of bureaucratic documents (Smith 1999: 152). Precisely because they draw their authority from feeling and not from objective reality, such texts are more open to expressing traumatic memory. LaCapra (2001: 41) says: "Being responsive to the traumatic experience of others, notably of victims, implies not the appropriation of their experience but what I would call empathic unsettlement, which should have stylistic effects or, more broadly, effects in writing which cannot be reduced to formulas or rules of method." Even in challenging the narrative of the inquiry report, all these writings, even my preface, have recourse to the findings of the inquiry report. Heffernan, Moore and I all start, in a sense, with that report. In fact, I cannot think of my brother's death, the *Ocean Ranger* disaster or even my adult life without that Royal Commission of inquiry.

THE POWER OF AFTERMATH REVISIONS

Each socio-political process performs its own kind of revision in the aftermath. An inquiry, with its testimony, cross-examination and presentation of evidence cultivates public confidence that factual truth is under the control of legitimate officials. The report of an inquiry makes a strong claim to accurately represent objective reality: a reality "out there" in the empirical past is represented in the report. As importantly, this reality is shown to be accessible through quasi-judicial evidentiary proceedings. Such reports frame memories of the traumatic event with exceptional power. In contrast to this, the power of a poem or a song like Ron Hynes' "Atlantic Blue" rests in a different kind of claim. The song's validity comes from its ability to evoke feeling: it invites us to feel grief safely as a gently melancholic pleasure. We are free to remember the sadness of the *Ocean Ranger* loss without being paralyzed by the shock of the first days.

In the aftermath of an industrial disaster the dominant version retells the event in terms that reinforce citizens' sense of passivity in the face of economic forces. As Adorno (1998 [1959]: 93) put it: "The dominant ideology of the day dictates that the more individuals are delivered over to objective constellations, over which they have, or believe they have, no power, the more they subjectivize this powerlessness." Many *Ocean Ranger* family members and former workers feel guilty that they did not know or did not do anything about the unsafe and brutal conditions on the rig. Like the underground workers at the deadly Westray coal mine, the men might have saved themselves by unionizing. Ray Hawco, the director of community relations for the Newfoundland and Labrador Petroleum Directorate believes they were moving in that direction before the storm overtook them. Nationalism, or even regionalism, can work to support governments when they assert the good of the local communities in the face of transnational capitalism, but this, too, is a fundamentally paternalistic model: we live out the destiny moulded for us by great men instead of participating in democratic deliberations of our own.

To challenge the forgetfulness offered by redemptive narratives where wrongs perpetrated by profit-seekers are recast as noble sacrifices is not only an act of personal remembrance. It is a political responsibility.

NOTES

1. The relationship between Americans and Newfoundlanders developed in the context of the American air force bases on Newfoundland soil from before the province's entry into Confederation: "The Yanks' were popular with the local population" (Earle 1998: 95).
2. "Judge Chaney... dismissed the suit in March, ruling that attorneys had recruited people who weren't banana farm workers as plaintiffs. According to the order, the attorneys coached the workers to lie, created bogus work certificates and lab results and also tried to intimidate witnesses and investigators. Sparks and co-counsel Benton Musselwhite were accused of taking part in the conspiracy, but the judge cleared them of charges that they perpetrated a fraud on the court or conspired to fabricate false testimony" ("Dole Offers Banana Workers Deal in Pesticide Row," June 15, 2011 <www.law360.com>).
3. C-45 added the following to the Criminal Code: "217.1 Every one who undertakes, or has the authority, to direct how another person does work or performs a task is under a legal duty to take reasonable steps to prevent bodily harm to that person, or any other person, arising from that work or task."

6

REMAKING THE PROMISE OF LIBERAL CAPITALISM: LEGITIMATION CRISIS AND WORKING THROUGH

"The tissues of the community can be damaged in much the same way as the tissues of the mind and body."

The ways we remember the *Ocean Ranger*, and the men we lost with it, are every bit as political as the regulatory failure that permitted the disaster. We make our own history but not in conditions of our own choosing, to return to Marx's famous saying. The landscape where we engage in political action is historical, just as we ourselves are historical. To begin to understand how liberal capitalism produces and reproduces the combination of freedom, exploitation and alienation that makes up the conditions in which we act, we need to know ourselves as historical beings. Liberal capitalism is characterized by ways of remembering and forgetting that reproduce the power of elites. These also offer opportunities for resistance: memory is political and it is unpredictable.

As a traumatic event, both for individual family members and for Newfoundland as a nation or an "imagined community," the disaster disturbed confidence in government and in the promise of oil, and it tore the very fabric of community life. Like hypertext links between levels of experience, trauma's triggers can transport a sufferer directly back to the moment of injury in repeated dreams, or it can disguise itself and capture the sufferer in nightmares, hallucinations or coded re-enactments that "act-out" the original injury.[1] The peculiar ability of an historical trauma to disorient us lies in its resonances with earlier, more primal or even symbolic injuries. The *Ocean Ranger* continues to have some potency as a public symbol because it links anxiety about economic development with old fears, the injuries from colonialism that Brian Peckford was determined to exorcize in the "new Newfoundland." Peckford's Newfoundland would escape from the regional inferiority complex that lay "deep in our psyche," resulting from Newfoundland's "whole history of colonialism, subjugation and exploitation" (Peckford, in Cadigan 2009: 267). The use of the *Ocean Ranger* disaster as a shorthand for government mishandling of the oil industry is still considered distasteful and disrespectful. In 2004, Nova Scotia Member of Parliament Gerald Keddy tried, during a debate over questions of Newfoundland and Nova Scotia getting 100 percent of their oil and gas revenues. "That offshore accord for Newfoundland and Nova Scotia is a greater tragedy than the *Ocean Ranger*," Keddy blustered (Postmedia News 2004). Even with a "successful" recovery, when the traumatic event is externalized in a story or work of art, we can be thrown right back to that first terrible moment. It all depends on the trigger. And triggers are idiosyncratic.

In contrast to long-term, community-poisoning collective traumas, such as the Holocaust or slavery, an industrial disaster restricts our gaze to a sudden event in a particular workplace. An industrial disaster breaks something, and the aftermath puts it back together but in a revised form. In studying the aftermath of an industrial disaster, we begin with an empirical shock that temporarily reveals a structural conflict. In the early hours, people are clear about the fundamental conflict at the heart of the disaster: the profit-motive of corporate activity does not naturally benefit host communities. In fact, when left without government regulation the drive to produce wealth for absent shareholders can only result in exploitation and destruction. By following the aftermath of an industrial disaster we can map the ways this conflict is then revised out of explicit public discourse.

Over the course of the disaster's aftermath, debates about blame and causation settled into a less emotional and less politically disruptive history. In this, the aftermath of the *Ocean Ranger* disaster offers an unusually clear picture of the politics of memory in liberal capitalism. In the forensic moments of the *Ocean Ranger* aftermath, the causes of this disaster were pieced together in such a way that confidence was restored in the development of offshore oil. As I have shown in previous chapters, and as I will show in the Conclusion in relation to the aftermath of the 2010 *Deepwater Horizon* explosion and oil spill, this kind of memory-making works through revisions, and so it has two sides: it heals and it hides.

To study the aftermath of an industrial disaster, or any disaster, is to study the relation between trauma and history. To work through a traumatic event is to give shape to feeling, to bring affect into a new relation with our words and actions. Trauma and working through are notoriously unpredictable. It is clear that individuals do not work through trauma at identical speeds. In fact, time does very strange things in the traumatized memory. The relation between the traumatic shock and the historical memory of a disaster is very complicated. To work seriously on the past and to avoid the easy fatalism or knee-jerk blame that can masquerade as genuine working through demands an interdisciplinary approach. As the ancient Greeks knew, memory—for them the goddess Mnemosyne—was the mother of all the muses. Art and in fact all human creativity originates in our working through the past. A traumatic event is worked through in many ways all at the same time as I have said, and all the diverse accounts exchange images, structures and even terms in an "intertextual" web (Caruth 1995; LaCapra 2001; Alexander et al. 2004; Adler et al. 2009).

REMEMBERING DISASTER: WRITING COLLECTIVE TRAUMA

Memories of the *Ocean Ranger* disaster are more complex than we like to admit, as memories tend to be. Stories about traumas are always more than literal representations of facts. Recovery from a trauma involves replacements: it is not that one dead person is replaced with a living one when a widow falls in love again, or that repeating dreams of loss are replaced with a yearly memorial service and a material monument. The process is far more complex and has multiple stages that work at different rates for different people, and at a different rate for the collectivity as an "imagined community." The stages of replacements or revisions overlap and coincide. One painful replacement that we have seen is when family members faced the evaluation of the dead man in cash terms: the financial equation of each man's life closes a phase of the recovery at a collective level by imposing terms that are experienced by some family members as a false closure, whereby they are somehow even symbolically implicated in the betrayal that lay at the heart of the initial loss. Another painful replacement occurred when the inquiry report stabilized a causal chain and concrete recommendations for policy reform to contain anxious speculation about what happened, who was to blame and what should be done. The inquiry testimony and expense exhausted public patience, and so the public, as an imagined community, were ready to write the report—to make the turn from forensics into action without dwelling on who was to blame. A liberating replacement for "the people of Newfoundland" is the novel *February*'s revision of the inquiry report's silence about the men's deaths: when Helen can finally "go there," she is able to distinguish her commitment to remember from a kind of living death.

Acknowledging the complex interplay between conscious action and unconscious symptoms that motivate people in the aftermath of a trauma has informed research and writing about the catastrophic events of the late twentieth and early twenty-first centuries. Sigmund Freud understood trauma to be a blow to the tissues of the mind that takes years to heal, if it heals at all. Trauma disorients us by separating our innermost feelings and psychic experiences from the things we say and do to express them: affect is dissociated from representation (Caruth 1995: 5). Freud "discovered" trauma first in studies of hysterical women, who he saw as having been "seduced" as children. He notoriously suppressed the implication that child abuse was at the root of these women's mental disturbances. Freud turned, then, in World War I, to study the strange effects of war injuries on soldiers: they relived their injuries over and over again, in flashback dreams and hallucinations. The Holocaust and the need to write its history, combined with the incapacity of ordinary description to represent the death camps, led a generation of thinkers to question the adequacy of mainstream historical writing. To gather the experiences of survivors into tidy stories was worse than inadequate; it repeated a kind of violence against the inexpressible experiences of the camps. In using existing categories for representing reality, all descriptions transformed the original event. Efforts to bear witness to the traumatic event can only produce revised versions, and so, in a sense, they erase and betray the memory (Edkins 2003). Historians and sociologists sought new kinds of writing to express the loss and suffering of trauma without laying claim to objective truth, that is, without restricting themselves to kinds of storytelling that demand demonstrations of evidence (White 2010; LaCapra 2001; Caruth 1995; Edkins 2003). Broadly conceived as a record of public memory, history includes many kinds of writings and commemorations. This is why we need an interdisciplinary approach if we want to follow Adorno's advice in aiming to break the past's power to control us.

The American sociologist Kai Erikson introduced the idea of "collective trauma" into sociological case studies of disaster (Erikson 1995). Erikson argues that a community can sustain an injury that strikes to "the basic tissues of social life," damages "the bonds attaching people together" and impairs "the prevailing sense of communality" (Erikson 1976: 154–54, in Alexander 2004: 4). Erikson (1995: 198) says:

> Human beings are surrounded by layers of trust, radiating out in concentric circles like the ripples in a pond. The experience of trauma, at its worst, can mean not only a loss of confidence in the self, but a loss of confidence in the surrounding tissue of family and community, in the structures of human government, in the larger logics by which humankind lives, in the ways of nature itself, and often (if this is really the final step in such a succession) in God.

For Erikson, a community is a "tissue" in which, properly, we encounter and recognize one another to create and recreate a common world. Collective trauma is, for Erikson, the loss of precisely this "We." As he puts it: "'We' no longer exist as a connected pair or as linked cells in a larger communal body" (Erikson 1976: 154, in Alexander 2004: 4).

An industrial disaster is a sudden loss of life and/or direct bodily injury sustained in a specific workplace. As a "disaster," the loss has a traumatic impact on the host community. It threatens to open public debate about the fundamental clash of interests between communities and profit-makers, and as a trauma, it "unsettles and forces us to rethink our notions of experience" (Caruth 1995: 4). A "disaster" problematizes or threatens the social fabric of a community: "One can speak of traumatized communities as distinct from assemblies of traumatized persons. Sometimes the tissues of the community can be damaged in much the same way as the tissues of the mind and body... but even where that does not happen, traumatic wounds inflicted on individuals can combine to create a mood, an ethos—a group culture, almost—that is different from (and more than) the sum of private wounds that make it up" (Erikson 1995: 185). A disaster like the *Ocean Ranger* loss is "industrial" in its workplace-specific location and its origins in the profit imperative, which prioritizes cutting costs over human and environmental vitality.

Sociologists have elaborated Erikson's insight into more fully developed methods for analyzing collective trauma. According to sociologist Jeffrey Alexander, imagined communities are constituted in the very activity of "constructing" a trauma, and consequently shock and betrayal lie at the heart of a community's self-image. Such social-construction theorists deny that there was ever a healthy community "before" the disaster. They argue that to believe in a healthy—even happy—organism that was self-aware and complete before it was shattered and confused by trauma is a "naturalistic fallacy." Some influential psychoanalytic accounts of collective trauma given in literary criticism and historiography are said to make the same mistake: "The truth about the experience is perceived, but only unconsciously. In effect, truth goes underground, and accurate memory and responsible action are its victims" (Alexander et al. 2004: 5).

By rejecting the idea that a *more* healthy community existed before, Alexander restricts his approach to a version of social construction theory. He conceives of Benedict Anderson's "imagined communities" as generated through the social construction of trauma. In this view, the event itself is unimportant as an historical moment. For Alexander and other constructivists, a cultural sociologist is a scientist who needs to be neutral about the accuracy of social actors' claims and moral justifications. "We are concerned only with how and under what conditions the claims are made, and with what results. It is neither ontology nor morality, but epistemology, with which we are concerned" (Alexander et al. 2004: 9). Alexander's ideal cultural sociologist seeks to describe, objectively, the practices by which a collectivity "constructs" a traumatic event, and even further, how such trauma constructions lie at the heart of collective identity.

In the radical constructivist view, the revisionary work of trauma constructs "a compelling framework of cultural classification" that responds to a number of questions, which I gloss as: What happened (and to whom)? Who is to blame? What is to be done? Alexander sees these questions as being addressed in one or a number of "institutional arenas," namely, religion, science, mass media and state bureaucracy. The dramaturgical, spatial metaphor of "arenas" captures the dynamic nature of the aftermath by presenting practices of revising affect into representation as performances in so many theatres. It fails, however, to consider these practices of collective working through as intimately interconnected and temporally unpredictable. This threatens to flatten the concept of trauma in adapting it from individual psyches to collective identity. The radical constructivist view also eliminates the prospect of distinguishing between kinds of accounts that are more rather than less adequate to the experience of loss and its effects. The temporal disruptions, latency, the triggers to more primary or existential senses of absence, distinctions between acting out and working through—much that makes trauma theory so rich an analytical approach is ignored.

In the previous chapter I presented the aftermath processes in a "map" of responses to a potential legitimation crisis. But a map is even more static than the "arenas" I complain about in Alexander's model, and so I want to complicate my image, to express the dynamic interplay between the reports, stories and works of art in which we work through a collective trauma. Let us now think of the aftermath of an industrial disaster as a web of interrelated styles of remembering that we can picture as a rope or a cable, but one erratically bound, frayed at points, tightly twisted at others, and always with loops where some strands come loose but then are wound back in. The inquiry produces a report, people talk and tell stories, the songwriters sing songs, and the novelists and oral historians combine strands from the inquiry report, folklore and songs when they write their books. Each strand of remembering spins its own kind of revision of the event, producing an account that makes sense because we recognize its kind of storytelling (its "genre"). Meanwhile, archivists and museum curators argue about what to throw out, what to preserve and how to balance their audience's desire for drama with "sound museological practice" (Babaian 2005: 72). In the aftermath of an industrial disaster, these all contribute to changing the focus or even to replacing the central issue: the conflict between the interests of communities and those of distant shareholders is replaced by a "history" of the event. This history includes a dominant version of the story, like the inquiry report, and versions that challenge the dominant style and content, as in Heffernan's *Rig* and Moore's *February*.

QUALITATIVE DISTINCTIONS BETWEEN WAYS OF REMEMBERING

My central aim is to describe but also to argue in favour of a sociological approach that admits explicitly qualitative distinctions between ways of representing the bad things that happen as a result of regulatory failure in liberal capitalism. We must both describe and make qualitative judgments about ways of remembering. Practices of describing the past differ in their roles in the aftermath, though they are always interrelated. Some are more open to traumatic memory while others are more objective or "factual" and limited by evidentiary requirements. Whereas it makes sense to restrict the state's use of violent punishment to strict evidentiary rules, the full historical expression of a traumatic event cannot be confined to "objectivist" writings.

Anthropologist Mary Douglas says: "The cultural dialogue is best studied in its forensic moments" (1990: 3). But it is not clear that we can neatly bracket forensic moments off from the rest of time. This is precisely the problem with traumatic memory. LaCapra argues:

> Trauma is a disruptive experience that disarticulates the self and creates holes in existence; it has belated effects that are controlled only with difficulty and perhaps never fully mastered. The study of traumatic events poses especially difficult problems in the representation and writing both for research and for any dialogic exchange with the past which acknowledges the claims it makes on people and relates it to the present and future. (2001: 41)

Was there really a healthy community before the disaster? Or was the "healthy community" posited as a lost object whose very construction constitutes collective identity? Is trauma an historical event resulting in the distortion of a comparatively healthy community, or is it an aspect of the human—or at least the modern—condition? I want to say yes to all these questions. A *more* healthy community existed before the traumatic event. A fantasy of a lost perfect community also plays a central role in the self-image of an "imagined community." This may be especially true in areas where community has resisted the mass culture exemplified by box stores, suburban sprawl and fast food chains. Communities struggling to sustain any character cling more or less self-consciously to a heritage. Generally speaking, "trauma" in its strange combination of rational self-awareness haunted by irrational flashbacks is an important way to describe Western culture after World War I.

In her reflections on the terrorist attack on New York City's World Trade Centre in 2001, Ann Kaplan describes her own experiences on the first days of that aftermath. As she wandered the streets of her city, she distinguished "the different levels through which the catastrophe was being 'managed'" (2005: 5). At the street level, Kaplan found a local collective trauma where groups of persons gathered and created spontaneous memorials. "Those rows of images of lost people overwhelmed me.... They made visible the need for closure, the awfulness of not knowing if a loved one is dead, and if dead, if one would ever have a body to mourn over" (5). At the same time, at the "levels of the media and discursive formations," Kaplan notes, "it gradually became clear that national ideology was hard at work shaping how the traumatic event was to be perceived." The dominant story of 9/11 was becoming part of America's self-identity (13). And at the "level of political identity and ruptures," Kaplan found herself divided from the self she felt she had been "before" 9/11. Like LaCapra, Kaplan seeks an account that is open to "the different modalities in which trauma is addressed" (22). And, she insists, politics and considerations of collective trauma are not mutually exclusive: it is possible to recognize the effects of trauma while still holding agents accountable.

Kaplan sees different "levels" and "modalities" managing the disaster's memory. In my view, there are also important exchanges between these levels and modalities: they borrow and revise from one another. At the street level, people engage in the communicative exchanges of ordinary life in order to reconstitute their community. At the level of national identity, the managerial, instrumental language of bureaucratic systems form their own versions, which will then be challenged in some regards and reinforced in others during the various judicial, quasi-judicial and financial processes that follow. According to sociologist Jurgen Habermas, the ways we solve problems when we are together in a room are more authentic and democratic than the ways bureaucrats and accountants solve them, *qua* bureaucrats and accountants. Administrative decision-making is instrumental, that is, its purpose is to render human action accountable in financial or other statistically measured terms.[2] Habermas calls the application of administrative categories to our daily lives the "colonization of the lifeworld" by systems instrumentality. When decision-making is replaced by conformity to statistical demands, the legitimacy of political and economic decision-making is no longer rooted in an embodied community (a lifeworld). Something is lost, and what once was at least responsive to the public becomes an abstract imperative to conform. This change is not just a shift, but a loss, even a degradation (Habermas 1986: 355; cf Habermas 1996: 342–52).[3]

Some of our ways of representing the world are more authentic than others. For Habermas, to try to understand each other is an exercise in good faith. To use language to manipulate each other breaches the logic of communication itself and is self-defeating, an act of bad faith. When we come to speak of ourselves and one another in terms of money and power, our practices of making sense of the world fragment, and thus ordinary life is "colonized" by the instrumental communication of the economic system (Warnke 1995: 121; Habermas 1986: 241). Administrative rationality imposes bureaucratic categories on people's lives. This entails the relentless and repetitive reduction of particulars to general categories and the evaluation of human life in monetary and statistical terms; when everything has a price, we are governed by an "identity thinking" that draws all values into market-monetary equivalence. The gradual restriction of families' demands in the aftermath of the *Ocean Ranger* disaster is an example of how far-reaching calls for justice and social change are narrowed into demands for money.[4] When people interpret their particular experiences in general terms with a view to maximizing profit or efficiency instead of coming to an understanding and seeking justice, community life is colonized by the administrative rationality of economic and political systems (Habermas, in Ingram 2010: 254).[5]

Habermas's distinction of administrative from ordinary modes of communication invites an interdisciplinary method for mapping disaster's aftermath. Because decision-making in day-to-day life takes place face-to-face in ordinary language, we can explore the pragmatics of symbolic interaction by using the tools of ethnomethodology: conversation analysis that is attentive to speech acts and impression management in their most minute details. We can also open spaces for memorial speech and writing and for recollections of experience that are not constrained by claims to empirical objectivity. Less constrained by narratives that participate in the "ideology of objectivism," this includes texts that challenge bureaucratic logic, including poems, interviews and discussions, art, oral histories, novels and songs. Lifeworld genres are negotiated, modified, revised and recreated by their participants.

Because decision-making in bureaucratic organizations conforms to the language of case files, memos and reports, we can explore the instrumentality of late capitalism by using the tools of literary criticism, particularly close readings or even more technical discourse analyses. The key is to follow the replacements of terms within any text, that is to say, the processes by which one practice of giving form to feeling is replaced by another practice of giving form to feeling. In ways similar to people's impression management, texts use literary figures to manage their reception by readers.[6] The tricks or revisions a text uses to draw the reader into it, as a collaborator in building the "facts" of the story, are called "tropes," literally "turnings." Harold Bloom (1975) goes so far as to call a trope "a willing error." The most famous trope may be metaphor, especially "love is a rose." Here, unrepresentable feeling is revised into a sensible object, a delicate, intricate, ephemeral flower with a flesh-ripping stem. Discourse analysts look to a text's techniques of impression management, and so they apply literary and linguistic criticism to inquiry reports, with the choice of analytical tools given by the text at hand. The text situates the reader while the reader interprets, and so, reading is a dialectical process.

Dorothy Smith (1999) holds that the reality construction of bureaucratic housekeeping is analogous to Marx's commodity fetishism: human activity becomes reified and turns back on its producers as a seemingly objective system. A person's problem becomes a case file, a case file becomes a statistic, a statistic becomes a bureaucratic issue, and the person with the problem is called upon to conform to bureaucratic specifications. Smith says a "fact is a mysterious thing, simply because in it the social character of men's consciousness appears to them as an objective character stamped upon the product of that consciousness" (68). Particular experience is rendered legible to bureaucratic categories and then those categories are turned back on the lifeworld. A bureaucratic document like an inquiry report conceals the traces of its own construction, readers should accept the report as unproblematically producing scientifically and administratively accessible "objects" out of embodied experience. A bureaucratic text conceals the human work that goes into creating its version of events: "Authorship is obliterated in the collaborative familiarity of 'we all know' theorising, or to put it more circumspectly, is reduced to secondary activities of collecting, classifying, editing, annotating, summarising, at the fringes of a text whose centre is filled by an objectively resistant reality beyond these words" (Green 1993: 111). In the same way that exposure disrupts impression management techniques (Goffman 1997), questions of authorship disturb the stabilizing fiction that a text is a transparent reproduction of material and situated reality. Discourse analysis can provide a kind of time-lapse photo of the colonization of community life by the instrumental logic of economics and poll-driven politics, and resistances to it.

Most interesting of all are moments when genres interact, when the ordinary language of community life is picked up and adapted by the administrative language of systems, and then turned back onto everyday life with a demand for conformity. This final turn, of systems instrumentality back into ordinary life, is administrative rationality at its most corrosive. This is the point when, as one Habermas scholar puts it: "Wherever average persons have been relieved of responsibility for problem solving in a particular area by transforming that area into a technical sphere of action, we can speak of independent systems whose actions constrain our everyday behaviour from the outside" (Ingram 2010: 254).

Where Kaplan sees "levels" at which the catastrophe is managed in relation to personal and collective identity, Alexander delineates institutional arenas—religion, science, mass media and state bureaucracy—which constitute cultural identity in the construction of a collective trauma (Alexander et al. 2004). Constructions of collective trauma are crucial to cultural identity, and they are closely related to the foundation myths of nationalism. For example, the American self-creation in a frontier land of rugged opportunity is both a triumph and a traumatic break from England. But historical traumatic events—like 9/11—also happen to "a people," and these resonate back to more original traumas in unpredictable ways. Conceiving of a social process as an arena, as Alexander does, presents kinds of memory-making as spaces of public appearance and contestation. Alternatively, conceiving of social processes as a web of interrelated strands of storytelling and the accounts they create points to the double movement of memory, that is, to the showing that is a concealing, the remembering that is a forgetting, as well as to the dynamic nature of the aftermath.

The aftermath is a web of versions of the story, each of which consists of a series of revisions, where one way of describing events and expressing emotions is replaced by another. In the previous chapter, I mapped out seven interrelated styles or genres of retelling the story that each expresses its own requirements for what counts as valid knowledge. Each kind of account entails a theory of community based on shared competencies: to be legible a text must meet the expectations of a reader who can recognize it for what it is and take up a way of interpreting that is appropriate to it. Genre is an underlying prescriptive code that gives speech or writing internal coherence so that it can make sense to listeners or readers. The account works between the text and the reader because they share communicative competency: the work makes sense between them. The plot of the story "'explains' not the events in the story but the story itself, by identifying it as a certain *kind* of story" (White 2010: 116). Representation uses symbols to point to empirical events or feelings. Each way of retelling an experience has its own kind of claim to truthfulness, some of which express what the literary theorist Hayden White calls the "ideology of objectivism" (313). For example, the inquiry report claims that it is a transparent window onto an empirically available past. A criminal trial would make an even stronger empirical claim, which is its ground for exacting punishment. Less "objectivist" texts are more reflective about their literary construction of "facts." White says that the important distinction "is not between ideology and objectivity but between ideological constructions of history that are more or less open about the 'constructed' nature of their version of history" (311).

RECOVERING FROM LEGITIMATION CRISIS

When disaster strikes, multiple interrelated socio-political processes kick into gear to help us heal but also to encourage us to forget, especially when the root cause of the disaster is the failure of governments to protect their public, let alone to successfully divert benefits from corporate activity into local communities. The *Ocean Ranger* disaster exposed a conflict between international corporations and the communities that host and staff their industrial operations, and this threatened to unleash a crisis: at the root of the disaster was governments' failure to regulate and corporations' prioritizing of profit over everything else. Laws that should have prevented the disaster were either non-existent or unenforced. For a short while at least, it seemed possible that the public might experience what the sociologist (Habermas 1976; 1984) calls a "legitimation crisis," that is, the belief that the economy offers no reasonable opportunities, and we cannot not trust the governments to protect our interests, or even our lives.

To work through a trauma is to give voice to the memory in an increasingly, though never entirely conscious way, so that the ghosts of past suffering become more and more companions, less and less demons that possess and mystify. Until we acknowledge and even welcome the ghosts, we are their captives, and they influence us in ways we refuse to admit, "acting out" in our self-loathing, rages, obsessions and anxieties. The healing is concrete once we are able to invoke the voices of the dead, instead of being commanded by them. To heal is to be able to distinguish legitimate demands for accountability from existential feelings of loss and alienation. Healing should make us more able to seek justice for historically specific wrongdoing without confusing it with our yearning to make God explain why His "plan" includes the untimely deaths of fathers, sons and friends (LaCapra 2001: 47–49). But healing is fragile, especially when our personal trauma originates in a betrayal of public trust that is repeated and exacerbated by a failure of justice to hold anyone accountable. Just when those of us traumatized by the sinking of the *Ocean Ranger* thought we had moved on, Cougar Flight 491 crashed on March 12, 2009, killing seventeen workers on their way out to the oilfields off Newfoundland's coast. Then, on April 20, 2010, the *Deepwater Horizon* blew up, killing eleven men and spewing oil for months into the Gulf of Mexico. For me, and I suspect for many friends and family of the *Ocean Ranger*'s crew, these disasters triggered flashbacks to that terrible night in 1982, and it left us wondering if, in our "healing," we had forgotten all the wrong things.

NOTES

1. While the precise definition of post-traumatic stress disorder is contested, most descriptions generally agree that there is a response, sometimes delayed, to an overwhelming event or events, which takes the form of repeated, intrusive hallucinations, dreams, thoughts or behaviours stemming from the event, along with numbing that may have begun during or after the experience, and possibly also increased arousal to (and avoidance of) stimuli recalling the event. This simple definitions belies a very peculiar fact: the pathology cannot be defined either by the event itself —which may or may not be catastrophic, and may not traumatize everyone equally —nor can it be defined in terms of a distortion of the event, achieving its haunting power as a result of distorting personal signifi-

cances attached to it. The pathology consists, rather, solely in the structure of its experience or reception: the event is not assimilated or experienced fully at the time, but only belatedly, in its repeated possession of the one who experiences it. To be traumatized is precisely to be possessed by an image or event. And thus the traumatic symptom cannot be interpreted, simply, as a distortion of reality, nor as the lending of unconscious meaning to a reality it wishes to ignore, nor as the repression of what once was wished (Caruth 1995: 4–5).

2. As unwieldy as Habermas's systems-lifeworld model is, it affords a qualitative distinction between ways of representing affect. Ordinary language is more adequate to human experience than bureaucratic language; and ordinary language is degraded when it is colonized by bureaucratic categories. In Habermas's formulation from the 1970s, a "legitimation crisis" threatens when the contradictions inherent in the economic system of capitalism are insufficiently mitigated by the welfare state. The key to sustaining the economic and political systems of late capitalism is the legitimation that the systems draw from decision-making of day-to-day life. Habermas's systems-lifeworld model is notoriously problematic: it gives too much credence to Parson's functionalism on one hand and it romanticizes existing community life on the other. Whatever the merits/limitations of these criticisms, Habermas's insight that decision-making operates according to multiple kinds of logic has been adapted by sociologists of language into a mixed-methodology discourse analysis that takes systems and lifeworlds as distinct discourses that interact in various genres of speaking and writing (Dorothy Smith, Bryan Green, Norman Fairclough).
3. Everyday consciousness sees itself thrown back on traditions whose claims to validity have already been suspended; where it does escape the spell of traditionalism, it is hopelessly splintered. In place of "false consciousness" we today have a "fragmented consciousness" that blocks enlightenment by the mechanism of reification. It is only with this that the conditions for a colonization of the lifeworld are met. When stripped of their ideological veils, the imperatives of autonomous subsystems make their way into the lifeworld from outside —like colonial masters coming into tribal society—and force a process of assimilation upon it (Habermas 1986: 355; cf Habermas 1996: 342–52).
4. "Habermas reinterprets Marx's base-superstructure distinction to establish the causal role of mechanisms operative on both sides of the distinction. Put simply, Habermas's appeals to institutions and the social processes that go on within them are necessary to explain how macro-to-micro relationships are established. Systems relate to institutions as superstructure to base: systems must be 'anchored' in the lifeworld through institutions (Habermas 1986: 259). The organization of the economy through market mechanisms requires that the systematically organized sphere of material production be 'anchored' in new cultural institutions such as civil law and bureaucratic offices. This interdependence makes 'increases in [system] complexity dependent on the structural differentiation of the lifeworld.' Furthermore, it is only through their anchoring institutions (and not some mysterious 'internal logic') that systems spread, as when social relations

become increasingly governed by law and exchange to define the character of social relationships (and not the intrinsic properties of the systems themselves) which has the dysfunctional-structural effects and the bad, unintended circumstances" (Bohman 1999: 75).

5. "Wherever average persons have been relieved of responsibility for problem solving in a particular area by transforming that area into a technical sphere of action, we can speak of independent systems whose actions constrain our everyday behaviour from the outside" (Ingram 2010: 254; Habermas 1986: 342–52).
6. Arguing for a "need for reflexive theorizing about textual reality construction," Green holds that "the need cannot be met adequately by focusing upon the concrete organizational practices through which records, documents, reports and so on are assembled and used, but only by extending to reading and writing the same close questioning of methods for accomplishing knowledge and reality effects which ethnomethodology has addressed to situated speech" (1993: 35).

Conclusion

The Promise of Oil? From the *Ocean Ranger* to the *Deepwater Horizon*

"If nothing else, this disaster should serve as a wake-up call."
—President Obama, May 27, 2010

The *Ocean Ranger* story is more than a sad story about a bad storm, tucked away in the Extreme Weather section of the CBC archives. While it has faded from memory internationally, in Newfoundland and to a lesser extent in the oil industry across Canada, the *Ocean Ranger* is a symbol of the bad old days before Newfoundland grew to maturity as an oil-producing province. Danny William's swagger about the vibrancy of the Newfoundland economy and its ability to benefit from offshore petroleum is unquestionably merited. Newfoundland is, after all, "dominating Atlantic economic growth" (CBC.ca June 3, 2011). Is this good luck the long-term pay-off from federal investments in oil exploration during the "heyday of the province's offshore oil and gas exploration" in the 1970s and 1980s (Baird 2011a)? Or is it some special prowess of the Newfoundland government?[1] I cannot say. Is it safer to work in the Canadian offshore now than it was in 1982? Yes, I believe it is, though the crash and deaths of Cougar Flight 491 raise new doubts. Any increased safety is due in equal parts to an improved regulatory structure, corporate attentiveness to the levels of risk the host community will tolerate and the presence of trade unions. Are Canadian waters and coastlines safe from ecological devastation, either from disasters like the 2010 Gulf oil spill or from the day-to-day spills that take place out of sight and out of mind of would-be regulators? Probably not. And what about the distribution of the economic benefits from the Canadian offshore? How far do the benefits really go?

When the *Deepwater Horizon* exploded on April 20, 2010, in the Gulf of Mexico, the earliest news stories told of the crew's successful evacuation from the blazing, melting rig. All but eleven men were safe. In the next twenty-four hours it became clear that the "missing" men were in fact dead. Over the next days the complicity of the regulators with negligent corporations grew more and more evident. Over the next weeks, oil gushed from the well. The Macondo well, named, incredibly, after Gabriel Garcia Marquez's fantastical town from

One Hundred Years of Solitude, was out of control. Even more incredibly, the oil companies resisted government attempts to determine just how much oil was spilling, and the world realized that no oil company has an effective plan to manage the massive oil spills that are not only possible but widely acknowledged to be probable (Safina 2011). BP, the owner of the rig, conducted what may have been one of the most awkward public relations campaigns in history. CEO Tony Hayward played a buffoon while the technological spectacle of attempting to staunch the flow of oil captured media attention. Meanwhile, here in Canada, Newfoundlanders anxious about the global impact of the oil spill headed out to see the gannets, sea birds that summer off our shores and winter in the Gulf of Mexico.

Five weeks after the *Deepwater Horizon* explosion, as oil continued to gush into the gulf, President Obama addressed America, emphasizing that he would decouple the government agency entrusted with approving drilling permits from the safety regulators and bureaucrats who were supposed to monitor disaster response plans. Responding to questions following this speech, Obama laid out the crisis of confidence in the governments' ability to control these massive companies that he knew loomed before him:

> Well, BP's interests are aligned with the public interest to the extent that they want to get this well capped. It's bad for their business. It's bad for their bottom line. They're going to be paying a lot of damages, and we'll be staying on them about that. So I think it's fair to say that they want this thing capped as badly as anybody does and they want to minimise the damage as much as they can. I think it is a legitimate concern to question whether BP's interests in being forthcoming about the extent of the damage is aligned with the public interest. I mean, their interests may be to minimise the damage, and to the extent that they have better information than anybody else, to not be fully forthcoming. So my attitude is we have to verify whatever it is they say about the damage. (Obama 2010)

The Gulf oil spill contributed to a growing sense of corporate profit-seeking uncontrolled by supposed government regulators, striking as it did amid the spectacle of financial predation ranging from subprime mortgages and foreclosures to the high-flying fraud of Bernie Madoff. As the American Congressional Financial Crisis Inquiry Commission found, regulatory "sentries were not at their posts" (*New York Times* 2011a). As the *New York Times* noted (2011b), the report of the Financial Crisis Inquiry Commission was not primarily a policymaking document but rather an ideological and historical one: "The report seems aimed at shaping future debate over the crisis. 'The greatest tragedy would be to accept the refrain that no one could have seen this coming and thus nothing could have been done,' the panel wrote in the report's conclusions. 'If we accept this notion, it will happen again.'"

The *Deepwater Horizon* explosion and Macondo oil spill were vivid demonstrations that when governments fail to regulate, corporate drive for profit leads to human, ecological and economic devastation. What had been obvious in the early hours of the *Ocean Ranger* aftermath was obvious once again: without regulation, profit wins out over community well-being, and catastrophe ensues. How did this happen again? How did it come as such a shock ... again? What happened to the "lessons learned" in the sacrifice of those eighty-four men in 1982?

Almost thirty years after the *Ocean Ranger* disaster, the kinds of sociopolitical processes that suppressed a potential legitimation crisis and secured Newfoundlanders' faith in the promise of oil in the early 1980s swung into high gear. The *Ocean Ranger* and the *Deepwater Horizon* are very different disasters, but they are identical in one crucial regard: in each case, governments anxious to harness petroleum profits for cultural renewal in a "have not" region left oil companies to operate without external regulation. In 1982, Canadians assumed that their governments would take reasonable measures to protect workers off the coast of Newfoundland; in 2010, Americans presupposed that their governments would do the same in the Gulf of Mexico. Both disasters showed citizens that they were wrong. The oil companies took freedom from regulation as an opportunity to cut costs and corners. In both instances, the trauma struck individual people and families, but it also struck the heart of "a people," a culturally distinct community characterized by a tradition of life on the sea, a history of entrenched poverty and a struggle for economic independence and recognition from the dominant culture. In both cases, the story would be transformed, with a major, even hegemonic, strand of the story presenting the disaster as a learning opportunity for technological engineering. As one of the few articles that considered the *Ocean Ranger* as a precursor to the *Deepwater Horizon* put it, disaster "can become a spur to innovation" (Broad 2010). That same *New York Times* article quotes Henry Petroski, a historian of engineering and the author of *Success through Failure*: "Nobody wants failures. But you also don't want to let a good crisis go to waste." The article continues:

> Another example in learning from disaster centers on an oil drilling rig called Ocean Ranger. In 1982, the rig, the world's largest, capsized and sank off Newfoundland in a fierce winter storm, killing all 84 crew members. The calamity is detailed in a 2001 book, "Inviting Disaster: Lessons from the Edge of Technology," by James R. Chiles.
>
> The floating rig, longer than a football field and 15 stories high, had

> eight hollow legs. At the bottom were giant pontoons that crewmen could fill with seawater or pump dry, raising the rig above the largest storm waves—in theory, at least.
>
> The night the rig capsized, the sea smashed in a glass porthole in the pontoon control room, soaking its electrical panel. Investigators found that the resulting short circuits began a cascade of failures and miscalculations that resulted in the rig's sinking.
>
> The lessons of the tragedy included remembering to shut watertight storm hatches over glass windows, buying all crew members insulated survival suits (about $450 each at the time) and rethinking aspects of rig architecture.
>
> "It was a terrible design," said Dr. Halada of the State University of New York. "But they learned from it." (Broad 2010)

That shutting storm hatches and providing survival suits on a vessel operating on the Grand Banks in February constitute "lessons of the tragedy" must be perversely funny to people who have lived and worked on those waters for generations. The article's author did not mention that someone had anticipated the problem of the ballast room's location and modified the *Ocean Ranger*'s sister rig, the *Dyvi Delta*. The *Dyvi Delta's* workers were lucky that someone in their rig's history learned the technical lesson in time to avert disaster. The real issue in both the *Ocean Ranger* and *Deepwater Horizon* cases, as President Obama clearly saw, was not a lack of technical know-how but a failure of regulation. The article notes that "numerous federal agencies are involved in a series of detailed investigations" of the *Deepwater Horizon* disaster, but that "the engineers hold, seemingly with one voice, that the investigatory findings will eventually improve the art of drilling for oil in deep waters—at least until the next unexpected tragedy, and the next lesson in making technology safer." The presupposition is that the root cause of industrial disaster is always a lack of technological know-how and never corporate neglect allowed by regulatory failure. As ODECO and Mobil did in the aftermath of the *Ocean Ranger*, BP did, and more to minimize the amount of money they paid out; this is not a moral claim, it is just what these organizations do. These corporations strategically control so-called "science," trace the cause of the event to a series of bizarre accidents leading to a single technical problem, suggest that worker error triggered the chain of bizarre events, manipulate local governments with the fear of lost revenue, jobs and in this case with the collapse of the whole industry, and finally claim to be chastened, having "learned a lesson" about something any reasonable and prudent person should have known in the first place (Dodd 1995, 2001).

If the problem really was a matter of learning from mistakes, then the

Ocean Ranger disaster would have been averted by lessons from the *Alexander Kielland*, which killed 123 men in the North Sea in 1980. That disaster should have taught governments and companies that they needed to invest in evacuation technology and training and to be more vigilant about safety regulation on rigs. The "lesson" had not been learned by the time my brother and his co-workers died just over a year after the *Alexander Kielland* loss. When the *Piper Alpha* exploded in 1988 off the coast of Scotland, it was close enough to shore that the horrific fire could be filmed: the world literally watched as the oil-fed fire consumed 167 men and melted the platform. Fifty-nine men lived, while the rest perished, some through following the "safety" protocol, which was based on the erroneous claim that the living quarters were fireproof. "In addition to regulatory change brought about by the *Ocean Ranger*, the British *Piper Alpha* disaster in 1988 influenced the Newfoundland safety regime, as it had in Norway. Provincial regulatory requirements were introduced or strengthened, especially those regarding permit to work systems, the layout and accommodation facilities of the rig, fire walls, and temporary safe refuges" (Hart 2005: 76). In the aftermath of each major disaster, industry and governments "discover" that whenever there is a conflict between safety and energy policy, safety loses.

When Gordon Jones died, at the age of twenty-eight, on the *Deepwater Horizon*, he left a young son and a wife who was expecting. He, like my brother Jim, was a "mud engineer." Gordon Jones' family became some of the most high-profile lobbyists in the early days of the *Deepwater Horizon* aftermath. Both Gordon's brother, Chris, and their father, Keith, are personal injury lawyers in Louisiana. As a family, Keith, Chris and Gordon's widow Michelle, decided to give every media interview requested of them. Chris told me in an interview that they knew they needed to keep public pressure on until the financial settlement was reached. Part of their lobby efforts targeted the legal anomaly by which the lawsuit for Gordon's death was restricted to projected lost wages. Under the *Death on the High Seas Act* (DOSHA), Americans who work on boats are limited in their claim for damages in ways that people who work on the land are not. They cannot sue for pain, suffering, for the loss of society or for "punitive damages," which are intended to punish those who are responsible for the loss. A National Public Radio story explained: "A court can compensate Ms Anderson [whose husband also died in the explosion] for direct economic loss for a death that occurs on the high seas, like her husband's estimated lifetime earnings, but not for the less tangible things like loss of care, comfort and companionship, as well as pain and suffering, that are a common part of suits concerning deaths on land" (Schwartz 2011).

The determination of Gordon Jones's family was evident in Christopher's presentation to the U.S. Senate Committee on Judiciary just more than two weeks after Gordon's death in the disaster. Chris argued: "Words cannot describe what Gordon meant to this family." He showed three slides, the first two of romantic family images haunted by Gordon's absence: a treehouse he was building for his oldest boy, and a family photo with mother, son and the newborn child that Gordon will never meet. Chris's third photo was of Gordon and his eldest son just after a golf lesson. Chris directly addressed BP CEO Tony Hayward: "He publicly stated he wants his life back. Well, Mr Hayward, I want my brother's life back.... We are asking you to amend DOSHA to allow for the recovery of non-pecuniary damages. Currently, Michelle, Stafford and Max can only recover pecuniary damages, comprised of Gordon's future lost wages minus income taxes and what Gordon would have consumed himself" (Jones 2010). Chris Jones argues that continuing to leave DOSHA in effect in offshore work should be thought of as a "drastic liability cap" for the industries it covers—cruise ships, ferries and oil rigs. To Jones, this means that those workers continue to be singled out for this cap, contrary to the U.S. commitment to "equal justice under law."

In an interview with *USA Today* about two weeks before the first anniversary of the *Deepwater Horizon* explosion, Gordon's widow, Michelle, describes ways that her recovery from the shock was impeded by the ongoing media coverage of the oil spill. The ecological and economic impact had "seemingly overshadowed the lives lost aboard the rig," the story explains and quotes Michelle: "We're grieving on a totally different scale.... There's no closure. There's no body. Did he know what was coming? Did he feel pain? And everywhere you go, there it is again" (Jervis 2011).

Keith Jones made Gordon's absence felt at the BP annual shareholders meeting around the same time. He wrote a letter that was read out to shareholders by an activist shareholder from the Gulf Coast, Antonia Juhasz. Jones's letter said: "This was no act of God—BP, Halliburton and Transocean could have prevented this. But it would have taken more time, more money, and you were too greedy to wait. You rolled the dice with my son's life, and you lost" (Hargreaves 2011). BP chief executive Bob Dudley responded to Juhasz' activist reading of the letter to the shareholders: "'Many of your statements sound like they were prepared by a plaintiff's attorney,' said Dudley, a comment that was also met with applause from the audience. Audience support for both Dudley and the woman underscored the difficulty BP has had in repairing its damage from the worst oil spill in U.S. history and getting on with the business of finding and producing more oil" (Hargreaves 2011). Gordon Jones's family used every available avenue to drive public attention to the men who died and away from the dominant story of technological evolution.

The media coverage of the race to cap the Macondo well became a technological melodrama that eclipsed even the heart-wrenching photographs of oil-covered birds. It pained family members, but it also gave them a sustained edge in their tort negotiations and lobbying. The day before the first anniversary of the *Deepwater Horizon* disaster and the death of his son Gordon, Keith Jones said: "I remember the day they capped that well—those images had been up in the corner of every TV screen, all that oil gushing into the ocean—I stopped seeing senators and congressmen and started seeing staffers" (Jones, in Baram 2011). The Jones family settled a few days before the first anniversary. "I can't say anything about the settlement, of course, but I can assure you that it did not come cheap," Chris Jones said. The *New York Times* reported: "Those familiar with the terms have said that the payments have reached as much as $20 million apiece" (Schwartz 2011).

The preface to the report from the National Commission on the BP *Deepwater Horizon* Oil Spill and Offshore Drilling shows that the Commission picks up President Obama's concern about a potential legitimation crisis. The report was released in January 2011 and preceded a number of further investigations, including two separate, parallel investigations that are ongoing at the Department of Justice, a criminal investigation and a civil investigation that involves government and private claims against BP, Transocean and others.

The title of the National Commission's report recasts the rig's name: *Deepwater Horizon* becomes simply, "DEEP WATER" in all caps, spread across an indecipherable photo of fire, smoke, water and twisted industrial wreckage. In this way the proper name of the rig becomes a sublime challenge. America is challenged, as a character, to venture bravely into the unknown, into the future. The report's subtitle signals the end of the forensic phase: "The Gulf Oil Disaster and the Future of Offshore Drilling." The report's author, the National Commission claims to allay a legitimation crisis:

> The disaster in the Gulf undermined public faith in the energy industry, government regulators, and even our own capability as a nation to respond to crises. It is our hope that a thorough and rigorous accounting, along with focused suggestions for reform, can begin the process of restoring confidence. (National Commission ... 2011: viii)

Dedicated to the dead men, the report expresses the remarkably modest "hope that this report will help minimize the chance of another such disaster ever happening again." In fact, the report is fraught with anxiety, and its concern with loss of control and "security" suggests that the shock of the explosion and spill resonates back to the attacks of 9/11. Ironic absence pervades the report: the eleven men are gone, the material evidence is inaccessible, the Gulf is polluted to an undetermined extent and the shocked nation has—temporarily—lost its capacity to act. The report explains: "A treasured American landscape, already battered and degraded from years of mismanagement, faced yet another blow as the oil spread and washed ashore. Five years after Hurricane Katrina, the nation was again transfixed, seemingly helpless, as this new tragedy unfolded in the Gulf." This report demands that all Americans share the risks of oil production in new ways: whenever you get in your car, you are responsible for the ongoing degradation of the Gulf Coast. But this, all of it, is presented as a kind of gift. As the report presents it, the loss was caused by freedom itself because where there is freedom, there is risk. Disaster is "technological advance." The report compares the oil disaster to the loss of the Columbia space shuttle, paraphrasing Charles Perrow's *Normal Accidents* thesis that, "'Complex Systems Almost Always Fail in Complex Ways." The commissioners, like the oil industry and the developers of the space shuttle, are presented as explorers: "The President said we were to follow the facts wherever they led." Though they were hindered by an absence of subpoena power, as they note in a teensy footnote: "The chief counsel's investigation was no doubt complicated by the lack of subpoena power. Nonetheless, Chief Counsel Bartlit did an extraordinary job building the record and interpreting what he learned. He used his considerable powers of persuasion along with other tools at his disposal to engage the involved companies in constructive and informative exchanges." The report challenges ordinary people in their daily lives—they should prepare to take their share of the increased risks of oil production. While the report acknowledges that improved government regulation and increased investment in research and development can replace the "lag" between explorations' ambitious striving and safety protocols, it presents the real task for the injured community as to recognize its own implication in the disaster. America must overcome its complacency towards petroleum development and shoulder its share of the necessary risk. The report initially presents the dead men as "crewmen" but by the end of the Preface, they are "fellow citizens" who lost their lives for the nation's fuel security.

Like the *Ocean Ranger* inquiry report, this *Deepwater Horizon* report attempts to impose a tidy map of shock, forensic work and then a clearly delineated turn to the future: this is the "desire to get on with things" that Adorno warned us about (1998 [1958]: 92). This falsified closure is a collective fantasy wherein the character "America" can simply choose to answer the call to action. What is needed, we are told, is a thorough rethinking of oil policy and our dependency on it in our daily lives. Once again, the failure of government to regulate triggered a collective trauma, a potential legitimation crisis that was managed by diverse socio-political processes. Again, in the aftermath of a disaster the event is revised to obscure the conflict between the injured community and the benefiting shareholders that lies at the root of corporate "complacency" about regulation. The inquiry process presents the forensic mode as temporary: we focus on the past for a limited term, settle on an account of what happened (and to whom), and who is to blame only in order to make a decisive turn to action. Then, the collectivity will be able to step out of the forensic mode into free creativity.

Tony Hayward, BP's chief executive, was dubbed "the most hated and clueless man in America" by the notorious British tabloid, the *Daily Mail*. In an article entitled, "BP oil Chief Captain Clueless lands new job in charge of health and safety," the *Daily Mail* (2011) reported that Hayward now stands to make millions in a position that puts him "in charge of environment, health and safety for commodities giant Glencore." Mere days following their deaths, media coverage mentioned the dead men infrequently, their loss eclipsed by the ecological catastrophe. Suddenly alert to the need for a good show of democratic control over the corporate risk-takers, President Obama announced: "I will not tolerate more finger-pointing or irresponsibility" and vowed to end the "cozy relationship" between oil companies and his federal regulators (*Chronicle Herald* 2010). It is disheartening to recall "the rather comfortable relationship between ODECO and the U.S. Coast Guard," which was discovered in the wake of the *Ocean Ranger* disaster almost thirty years before. As Karl Nehring, a one-time captain of the *Ocean Ranger* who quit because ODECO and Mobil would not clarify that he had charge of the rig when it was not drilling, put it, "The oil companies are so powerful that they can ride the United States Coast Guard. That's my opinion" (in O'Neill no date: 25).

Following the *Deepwater Horizon* disaster, the scale of BP's crisis seemed at first to outstrip other oil disasters in relatively developed parts of the world. The disaster and ensuing finger-pointing between BP and Transocean conveyed so clearly that the profit imperative trumps not only the health and safety of people who work for such corporations but the health of people who work for them but the health of the human and natural community as a whole. President Obama's announcement that he would end the "cozy relationship" between regulators and the oil industry was an essential part of the "learning story." The president needed to show that he could not have seen foreseen the problem, that he learned from the catastrophe, held the culprits accountable and took decisive steps to change things for the future. The problem is that the level of change that the *Deepwater Horizon* explosion and Gulf oil spill demands is revolutionary: to eliminate offshore drilling or to institute policies for corporate polluters that are tough enough to kill the offending incorporated persons.

Despite early speculation about whether BP would survive the expense of the clean-up and litigation, "BP's stock has recovered considerably since the huge hit it took following last year's disaster.... BP's shares lost over half their value in the weeks following the spill, going from over $60 a share to under $30 in two weeks. The drop even prompted talk that the world's fourth-largest company could become a takeover target. They now trade around $45 a share, up more than 50% from lows last May" (Hargreaves 2011). There was even better news for BP, as it continued its public squabble with Transocean over responsibility for the disaster: on July 26, CNN quoted new BP CEO Bob Dudley: "It was only a year ago—just—that oil was still flowing into the Gulf... We've re-stabilized the company, re-strengthened the balance sheet. We've announced new exploration deals across the globe, our credit ratings have increased, and we're back heading in the directions we need to" (B. Jones 2011). BP paid out $6.8 billion as a result of the oil spill and now reports a $5.3 billion profit.

The *Deepwater Horizon* story is still taking shape, and the disaster has played out the aftermath pattern I tracked in the *Ocean Ranger* with sickening accuracy. All of the elements at work in Newfoundland in 1982 are again in play, but on a colossal scale. By now it is a cliché to note that corporations deploy a consistent range of memory strategies in the aftermath of industrial disaster, and BP reported a $3.7 million dollar investment in lobbying in the first half of 2011.[2] From Newfoundland to the Gulf of Mexico, and around the world, corporate agents problematize causal chains that lead back to the corporate person, emphasizing those that lead to particular workers instead. Transocean's and BP's antics in the early days of the *Deepwater Horizon* catastrophe are textbook in that regard.

To compare the oil companies' use of power to control information in the aftermath of the *Ocean Ranger* and the *Deepwater Horizon* losses is to compare across jurisdictions and across time. In 1982, in Newfoundland, Justice Hickman claimed the evidence from the ocean floor and had the Mounties take charge of it (Hickman 2008). In 2010, in Louisiana, BP and Transocean controlled access to videos of the spewing oil well, limited media and public access to the very beaches during the clean-up and imposed a tightly disciplined message on the twelve-dollar-an-hour workforce cleaning the tar balls, mousse and dead animals from the beaches and wetlands. They were told to respond "no comment" to any media or public questions. State and municipal governments received direct payments from Transocean as they developed their own ad hoc responses to the oil washing up on their beaches and in their wetlands. In the *Deepwater Horizon* case, testimony obviously repeated talking points. While BP was famously ham-handed with its public relations, it did deploy a public relations strategy in a way unheard of in the early 1980s (Safina 2011). The institutional culture of the oil companies appears to be relatively unchanged, or even more profoundly independent of the interests of local populations and environments. The contest over the story of catastrophe had changed dramatically in those intervening decades. ODECO's highest ranking workers testified at the *Ocean Ranger* inquiry on the condition that their appearances not be televised. ODECO used all its legal means to cast the story in a way that deflected attention from its own institutional negligence, but it did so in a way that left Justice Hickman respectful of its lead lawyer's jurisprudence. In the *Deepwater Horizon*, top figures pled the Fifth Amendment, refusing to respond to questions on the grounds that this testimony might be used against them in future criminal charges. In the 1980s, in Canada, corporate control of information was more subtle, and it seems to have opened the inquiry to a thoroughgoing exploration of "What happened? Who was to blame? and What should be done?" The *Ocean Ranger* was at the beginning of Canada's offshore career; Newfoundlanders were still being convinced by their politicians and other community leaders that oil was the promise of the future. In Louisiana, in 2010, the forensic cultural dialogue was grounded in the presupposition that oil work was already a way of life, a "tradition" almost as deeply entrenched as fishing.

On the first anniversary of the *Deepwater Horizon* disaster, Gordon Jones's family held their own quiet memorial, they played golf—Gordon's favourite sport—had a crawfish boil and drank Gordon's favourite beer. Their lives are marked forever by their loss and by their shaken confidence in their governments' ability to ensure that oil companies work *for* the people, and not at their expense. This kind of loss can also be an obligation to the future. Chris Jones expressed exactly what I felt as an *Ocean Ranger* family member when I heard about the *Deepwater Horizon*:

> The next time disaster happens, I'm going to be there, and when they're lobbying for these changes, I'm going to be there. Because I told every legislator when I walked in their offices ... and I told every one: "This is going to happen again. And people are going to be in your office again saying the same thing." And I know I'm going to be there with whoever. (C. Jones 2011)

Corporate self-regulation is a myth. This myth is exposed, as such, whenever corporate risk-taking suddenly and obviously injures a community. The *Ocean Ranger* disaster of 1982, the *Deepwater Horizon* disaster of 2010, and the global economic crisis triggered by the financial fraud and mismanagement in the new millennium are all legitimation crises: that is, they are obvious failures of our political and economic systems to work for the good of our communities. These legitimation crises and the collective trauma they inflict are managed through a web of socio-political processes, as we saw in the aftermath of the *Ocean Ranger* disaster. These processes work together to meet some demands of justice, to shed light on some truths, and also to impose a false closure that limits public reflection on gross regulatory failures.

As my father sees it: "We, as citizens in a democracy, have a responsibility to ensure that our various levels of government enact tough rules to protect our workers and our environment, and that those rules are enforced because corporations are in the business of making money and they will cut corners." I do not always share his confidence that the largest corporations in the world can be made to answer to us through our governments. The issue is not one of financial and legal power alone. It is also a question of memory, of how stories are told, where, by whom and to whom. To maximize profit, corporations expose us to exactly the level of risk they think their host communities will tolerate. This is a simple and obvious point and yet corporate self-regulation is a resilient myth. The eighty-four men on the *Ocean Ranger* did not need to die as a sacrifice on Newfoundland's road to maturity as an oil-producing province any more than the *Deepwater Horizon* catastrophe was needed to teach Gulf Coast regulators to demand spill management plans from companies. The "learning story" of disaster's aftermath is a myth that resurrects the myth of corporate self-regulation. It reestablishes the presupposition that corporations are naturally good citizens, experts in their fields, and so they should be left to develop resources and know-how for the general good. Time and time again, this is proven to be false. Time and time again, our ways of capturing these disasters in history narrow the memory of causation from complex socio-political crisis to a simple chain of technical mishaps or personal immorality. To argue for government regulation is only one way to respond to exploitation and inequality. To show the diverse ways that our economic and political systems use our own loves and values against us to shore up their apparent legitimacy is another approach. Our responsibilities span our diverse roles, as policymakers, activists, students, teachers, artists, writers, family members, singers, gossips and historians. How we tell the tale is literally what we make of ourselves as a community.

NOTES

1. Perhaps exploration is down because "it's a harsh environment and you've got high regulatory requirements," as some hold (Baird 2011b).
2. "BP… has so far spent 12 percent more on lobbying this year that it did during the first half of last year, $3.7 million versus $3.2 million. Much of the company's lobbying occurred in the wake of the *Deepwater Horizon* oil spill last April. This sum, however, is 51 percent less than its lobbying expenditures during the first half of 2009 when it used a broader disclosure method" (July 25, 2011, "High-energy companies reduce lobbying expenditures," *Open Secrets Blog: <http://www.opensecrets.org/news/2011/07/energy-companies-reduce-lobbying.html>*).

SELECT BIBLIOGRAPHY

ABA *Journal*. 1986. "Benton Musselwhite." Vol 72, Feb 1: 42.

Adler, Nancy, et al. 2009. *Memories of Mass Repression: Narrating Life Stories in the Aftermath of Atrocity*. London: Transactions Publishers.

Adorno, Theodor. 1998 [1959]. "The Meaning of Working Through the Past." *Critical Models: Interventions and Catchwords*. New York: Columbia Press.

ADR Chambers International. 2011. "The Honourable T. Alexander Hickman, Q.C." <www.adrchambersinternational.com/cvhickman.htm>.

Alberta Report. 1982. "One Thing Those Bereaved Have a Right to Know: A Letter from the Publisher [Ted Byfield]." March 1.

Alexander, Jeffrey, et al. 2004. *Cultural Trauma and Collective Identity*. Berkeley: University of California Press.

Amherst Daily News. 1983. "Oil Moratorium Sought." March 4.

Anderson, Benedict. 2003 [1983]. *Imagined Communities*. London: Verso.

Anderson, Patricia. 1980. "Newfoundland Fumes at Ottawa." *Financial Post*, May 24.

Antle, Rob. 2002a. "Ocean Ranger: As the Years Passed, Fallout from the Tragedy Spread." *Western Star*, February 16.

____. 2002b. "Ocean Ranger: 20 Years Later, Offshore Safety Is Still a Concern for Some." *Western Star*, February 18.

____. 2002c. "Ocean Ranger: Time Hasn't Diminished Memories of Those Lost." *Western Star*, September 6.

Auden, W.H. 1946. "Under Which Lyre: A Reactionary Tract for the Times." in *Selected Poems*. New York: Vintage Books.

Babaian, Sharon. 2005. "Evidence From a Disaster: The Ocean Ranger Collection at the Canada Science and Technology Museum." *Material History Review* 61: 69–78.

Bailey, Sue. 2011. "Nfld Premier Disputes Federal Air Rescue Centre Closure." Politics.ca, June 21.

Baird, Moira. 2011. "Exploration Glory Days." *Telegram*, April 29.

____. 2011b. "Exploration in Decline." *Telegram* April 30.

____. 2011c. "Turning the Exploration Tide." *Telegram* May 2.

Baker, Wallace. 1983. "Church Packed for Ocean Ranger Service." *Leader Post*, February 16.

Baram, Marcus. 2011. "Nil, Baby, Nil: Congress Fails to Pass a Single Oil Spill Law." *Huffington Post*, May 19.

Barmash, Pamela. 2004. "Blood Feud and State Control: Differing Legal Institutions for the Remedy of Homicide During the Second and First Millennia B.C.E." *Journal of Near Eastern Studies* 63, 3 (July): 183–99.

Barry, Leo. 1982. Letter to Noreen O'Neill. March 1.

BBC News online. 2010. "Obama: 'No More Cosy Relationship.'" May 14.

Beck, Ulrich. 1992. *Risk Society: Towards a New Modernity*. London: Sage.

Bennet, W.T., and J.L. Goldman. 1983. "Strength and Stability of Mobile Drilling Units." Safety of Life Offshore: Proceedings from the Symposium June 8–10 International Association of Drilling Contractors and Scripps Institution of Oceanography. La Jolla California.

Berman, Herald T. 1978. "The Background of the Western Legal Tradition in the Folklaw of the Peoples of Europe." *University of Chicago Law Review* 45, 3 (Spring): 553–97.

Bewes, Timothy. 2002. *Reification or the Anxiety of Late Capitalism*. London: Verso.

Bloom, Harold. 1975. *A Map of Misreading*. Oxford: Oxford University Press.

Bohman, James. 1999. "Habermas, Marxism and Social Theory." In Dews (ed.), *Habermas: A Critical Reader*. Oxford: Balckwell.

Broad, William. 2010. "Taking Lessons From What Went Wrong." *New York Times*, June 10.

Bronstad, Amanda. 2011. "Dole Settles Pesticide Claims With 5,000 Former Banana Plantation Workers." *Texas Lawyer*, June 15.

Buchan, David. 1982. "Propp's Tale Role and a Ballad Repertoire." *Journal of American Folklore* 95, 376 (Apr.–Jun.): 159–72).

____. 1991. "'Sweet William's Questions': A Report on Continuing Research into 'Treason Songs.'" In Thomas and Widdowson (eds.), *Studies in Newfoundland Folklore: Community and Process*. Toronto: Breakwater.

Bursey, Brian. 2010. Interview June 10.

Cadigan, Sean. 2009. *Newfoundland and Labrador: A History*. Toronto: University of Toronto Press.

Calgary Herald. 1983a. "Ocean Ranger: Solemn Service Replaces Legal Battling." February 11.

____. 1983b. "Rig Suits Stall." February 25.

______1983c. "Suits Settled for Ocean Ranger Families." December 29.

Campbell, Bill. 2011. "Regulatory Failures: From Ocean Ranger to Deepwater Horizon and What Wasn't Learned." February 21. Oil and Gas IQ <www.oilandgasiq.com>.

Campbell, Mary, and Susan Dodd. 1993. "Lessons of Disaster." *New Maritimes* 11, 5 (May/June): 17–19.

Campbell, Wayne. 1984. "The Legacy of the Ocean Ranger." National Research Council Canada. *Science Dimension* 16, 6: 8–17.

Canada. 1986a. "Federal Government Response to the Report of the Ocean Ranger Royal Commission," Press Line.

Canada. 1986b. "Minister Appoints Ocean Ranger Task Force." Press release, April 14.

Carson, W.G. 1984. *The Other Price of Britain's Oil: Safety and Control in the North Sea*. Oxford: Martin Robertson.

Cape Breton Post. 1984. "Operators Never Told How to Use Backup System." March 23.

____. 1983. "Hiring Local Workers Over Veterans Caused Training Problems Says Boss." March 16.

Carter, Glen. 2006. "Call in the Night." *Newfoundland Herald*. February 19.

Caruth, Cathy. 1996. *Unclaimed Experience: Trauma, Narrative, and History*. Baltimore: Johns Hopkins University Press.

____. 1995. *Trauma: Explorations in Memory*. Baltimore: John's Hopkins University Press.

Cassells, Jamie. 2008. *Remedies: The Law of Damages*. Toronto: Irwin Law.

CBC. 2011. "N.L. Dominating Atlantic Growth: Forecast." June 3.

Chronicle Herald. 2010. May 15: B5.

____. 1982. "Ocean Ranger Families Foundation: The Fight Is Just Beginning." July 12.

Clugston, Michael. 1982. "An Aftermath of Sorrow and Anger." *Macleans* March 1.

CNN. 2011. "Gulf Oil Rig Owner Apologizes for Calling 2010 'Best Year' Ever." April 22.

Cowell, John. 1701. *The interpreter of words and terms, used either in the common or statute laws of this realm, and in tenures and jocular customs*. London: W. Battersby.

Crosbie, Chesley. 2011. Interview July 21.

____. 2004. "The Newfoundland and Labrador Class Action Experience and the Risks of Jurisdictional Competition." <www.chescrosbie.com>.

Crosbie, John. 2008. Interview December 16.

____. 2007. "Criminal Inaction." *Independent* 5, 29 (January 20).

____. 2003. "Overview Paper on the 1985 Canada-Newfoundland Atlantic Accord." Discussion paper for the Newfoundland and Labrador Royal Commission on Renewing and Strengthening Our Place in Canada, the Grimes Commission.

Daily Mail. 2011. "BP Chief Blamed for Gulf of Mexico Oil Disaster Lands a Lucrative New Job… in Charge of Environment, Health and Safety for Commodities Giant Glencore." May 12.

Daily News. 1983. "Offshore Riches Carry Awesome Burden for Islanders." February 8.

____. 1982. "Ocean Ranger Families Foundation Asks Mobil to Suspend Drilling." March 10.

Daly, Martin, and Margo Wilson. 1988. *Homicide*. Aldine de Gruyter: New York.

Dodd, Susan. 2006. "Blame and Causation in the Aftermath of Industrial Disasters: Nova Scotia's Coal Mines from 1858 to Westray." In E. Tucker (ed.), *Working Disasters: The Politics of Recognition and Response*. New York: Baywood Press.

____. 2001. "The Writing of 'The Westray Story': A Critical Discourse Analysis." PhD thesis, York University.

____. 1995. "Restitching Reality: How TNCs Evade Accountability for Industrial Disasters." *Alternate Routes: A Journal of Critical Social Research* 12: 23–61.

Douglas, Mary. 1990. "Risk as a Forensic Resource." *Daedalus* 119, 4 (Fall).

Doyle, Pat. 1982. "Ottawa Considers Province's Offer for Joint Inquiry into Oil Rig Sinking." *Evening Telegram*, February 23.

Doyle, Pat. 1982b. "The Search Will Continue: Ships Carrying Bodies Return Today." *Evening Telegram* February 19.

Dunn, Sheilagh M. 1981. *The Year in Review, 1982: Intergovernmental Relations in Canada*. Kingston, ON: Institute of Intergovernmental Relations, Queen's University.

Earle, Neil. 1998. "Survival as a Lively Art: The Newfoundland and Labrador Experience in Anniversary Retrospective." *Journal of Canadian Studies* 33, 1 (Spring): 88–107.

East Coast Petroleum Operators' Association. 1983. Annual Report.
Edkins, Jenny. 2003. *Trauma and the Memory of Politics*. Cambridge: Cambridge University Press.
Erikson, Kai. 1995. "Notes on Trauma and Community." In Caruth (ed.), *Trauma: Explorations in Memory*. Baltimore: Johns Hopkins Press.
Evening Telegram. 1990. "School Remembers Ocean Ranger with Memorial Mass Thursday." February 14.
____. 1985. "Parents of Ranger Victims Can't File Suits in U.S." August 17.
____. 1984a. "Doctor Suggests Many See Drowning as 'Act of God.'" August 22.
____. 1984b "Skeletons Not Recovered." August 30.
____. 1984c. "No Chance Left Now Skeletons to Be Found." October 1.
____. 1983. "Memorial Service Near Wreck of Ranger." August 20.
____. 1983b. "Ranger finally at rest." August 27.
____. 1982a. "Next of Kin Cautioned by Ottenheimer." February 18.
____. 1982b. "Mobil Won't Comment on Charges by Former Skipper of the Ranger." February 19.
____. 1982c "Lawsuit filed." February 19.
____. 1982d. "Ottawa Considers Province's Offer for Joint Inquiry into Oil Rig Sinking." February 23.
____. 1982e. "RCMP Probes Rig Deaths." February 25.
____. 1982f. "Ocean Ranger Suit Filed." March 20.
____. 1982g "Crosbie Says Offshore Dispute Make-Or-Break Issue for Province." April 2.
____. 1982h. "Provincial Voters Offered Choice Between Fighter and Conciliator." April 5.
____. 1982i. "Chairman Gives Ultimatum to ODECO on Testifying." November 6.
____. 1982j. "ODECO Will Permit Employees to Testify." November 18.
____. 1982k. "Editorial." July 17.
____. 1982l. "Ocean Ranger Foundation Set Up." April 27.
____. 1982m. April 3.
____. 1982n. "Rig Goes Down; Bodies Sighted. 84 People on Board." February 15.
____. 1982o. "Safety Procedures on Rig Criticized." February 16.
____. 1982p. "The Perils of the Sea." February 16.
____. 1982q. "High Wind and Turbulent Sea Prevented Recovery of Bodies." February 16.
____. 1982r. "Chronology of Oil-Rig Disaster." February 16.
____. 1982s. "Sorrow Expressed by Government." February 16.
____. 1982t. "Diver Amazed Rig Went Down." February 16.
____. 1982u. "Mobil Will Supply Vessel to Search for Ocean Ranger." February 16.
____. 1982v. "Rig Deaths Total 169." February 16.
____. 1982w. "T. Alex Hickman: Man Who Heads Federal Inquiry." February 19.
____. 1982x. "Ottawa Orders Flags Flown at Half-Mast." February 19.
____. 1982y. "Security Tight at Mobil Office." February 19.
____. 1982z. "Province Mourning Rig Victims." February 19.
____. 1982aa. "Compensation Board Says: Owners Can't Be Sued." February 19.

___. 1982ab. "The Telegram's Statement on the Ocean Ranger Fund." July 17.
___. 1982ac. "United States Launches Inquiry Into Sinking of Ocean Ranger." April 27.
___. 1982ad. "Ocean Ranger Inquiry." April 22.
___. 1982ae. "Rig Compensation: U.S. Places Bill on Hold." September 30.
___. 1982af. "Ocean Ranger Monument to be Simple, Tasteful." October 20.
___. 1982ag. "Winter Drilling Program Approved by Government." November 5.
___. 1982ah. "Drilling in Winter." November 9.
___. 1982ai. "Premier Wants Immediate Action on Search, Rescue Improvements." December 23.
___. 1982aj. "Ranger Scholarship Established." August 10.
Fiss, Owen M. 1984. "Against Settlement." *Yale Law Journal* 93.
Flood, Mary. 1997. "What Did He Do to Deserve This?" *Houston Press*, January 23
Flower, Harriet. 2006. *The Art of Forgetting: Disgrace and Oblivion in Roman Political Culture*. Chapel Hill: The University of North Carolina Press.
Fortun, Kim. 2001. *Advocacy after Bhopal: Environmentalism, Disaster, New Global Orders*. Chicago: University of Chicago Press.
Fredericton Gleaner. 1983. "Ocean Ranger Disaster: Management Failure Blamed." February 9.
Freud, Sigmund. 1989 [1917]. "Mourning and Melancholia." In Peter Gay (ed.), *The Freud Reader*. New York: W.W. Norton.
Gerbeau, Samantha. June 10, 2010. Interview.
Globe and Mail. 1983. "Ocean Ranger Report Blames Rig Owner." February 9.
___. 1982a. "Rig Owners Silent on MP's charge." November 3.
___. 1982b. "Sneak Move Will Bar Claims by Rig Workers." December 14.
___. 1982c. "Rig Victims' Kin Offered Pact." December 11.
Godsoe, J.G. 1990. "Comment on Inquiry Management." *Dalhousie Law Journal* 12, 3: 71–73.
Goffman, Erving. 1997. *The Goffman Reader.* (Lemert and Branaman, eds.) Cambridge, MA: Blackwell.
Gold, Edgar. 2011. Email July 25.
Goldsmith. 2009. Damages for Personal Injury and Death in Canada 2002–2009.
Gordon, Steve. 2010. Interview April 21.
Government of Canada. 1986. "Minister Appoints Ocean Ranger Task Force." April 14.
Green, Bryan S. 1993. *Gerontology and the Construction of Old Age: A Study of Discourse Analysis*. New York: deGruyter.
___. 1982. *Knowing the Poor: A Case-Study in Textual Reality Construction*. Boston: Routledge & Kegan Paul.
Gzowski Films. 2002. "The Ocean Ranger." Film documentary.
Habermas, Jurgen. 2008. *Between Naturalism and Religion: Philosophical Essays*. (C. Cronin, trans.). Cambridge: Polity Press.
___. 1996. *Between Facts and Norms: Contributions to a Discourse Theory of Law and Democracy*. (W. Rehg, trans.). Cambridge: MIT Press.
___ 1986. *The Theory of Communicative Action. Volume 2: Lifeworld and System: A Critique of Functionalist Reason*. (McCarthy, trans.). Boston: Beacon Press.

____. 1984. *The Theory of Communicative Action. Volume 1: Reason and the Rationalization of Society.* (McCarthy, trans.). Boston: Beacon Press.
____. 1976. *Legitimation Crisis.* (McCarthy, trans.). London: Heinemann.
Hacking, Ian. 1990. *The Taming of Chance.* Cambridge: Cambridge University Press.
Halifax Mail Star 1983. "Safety Sole Concern." March 4.
Halpert, Violetta Maloney. 1991. "Death Warnings in Newfoundland Oral Tradition." In Thomas and Widdowson (eds.), *Studies in Newfoundland Folklore: Community and Process.* Toronto: Breakwater.
Hargreaves, Steve. 2011. "BP Victim's Father: 'You Rolled the Dice and Lost." CNN Money.com, April 14
Harries, Jill. 2007. *Law and Crime in the Roman World.* Cambridge: Cambridge University Press.
Hart, Sue. 2005. "Safety and Industrial Relations in the Newfoundland Offshore Oil Industry Since the Ocean Ranger Disaster in 1982." In Charles Woolfson and Matthias Beck (eds.), *Corporate Social Responsibility: Failures in the Oil Industry.* New York: Baywood Publishing.
Hawco, Ray. December 2008. Interview.
____. 1983. "Response to Human Disaster: The Ocean Ranger." *Canadian Journal of Community Mental Health.* Special Supplement, No 1: "Psycho-Social Impacts of Resource Development in Canada: Research Strategies and Applications."
Hayashi, Tosh. 1983. "Offshore Casualties." *Alberta Law Review* 21
Heffernan, Mike. 2009. *Rig: An Oral History of the Ocean Ranger Disaster.* St John's: Creative Publishers.
Heising, Carolyn D., and William S. Grenzebach. 1989. "The Ocean Ranger Oil Rig Disaster: A Risk Analysis." *Risk Anaysis* 9, 1.
Helliwell, John, Mary MacGregor, Robert McRae, and Andre Plourde. 1989. *Oil and Gas in Canada: The Effects of Domestic Policies and World Events.* Toronto: Canadian Tax Foundation Publications.
"Here and Now." 1982–84. Quotations from a CBC library catalogue of emissions.
Hickey, Patricia. 1984. Interview conducted by Dr Douglas House.
Hickman, T. Alex. 2008. Interview December 12.
____. 1983. "Statement by Chief Justice, T. Alex Hickman, Chairman of the Royal Commission on the Ocean Ranger Marine Disaster at the Conclusion of the Second Phase of the Commission's Hearings." June 2.
Hiller, James K. 2008. "Epilogue." In Newfoundland Historical Society (ed.), *A Short History of Newfoundland and Labrador.* Portugal Cove-St Phillips: Boulder Publications.
Holmes, Brooke. 2007. "The Iliad's Economy of Pain." *Transactions of the American Philological Association* 137, 1 (Spring): 45–84.
Homer. 1974. *The Iliad.* (Fitzgerald, trans.). New York: Doubleday.
House, J.D. 1987. *But Who Cares Now? The Tragedy of the Ocean Ranger.* St. John's: Breakwater Press.
House, Douglas. 1985a. *Working Offshore: The Other Price of Newfoundland's Oil.* Canadian Petroleum Directorate.

____. 1985b. *The Challenge of Oil: Newfoundland's Quest for Controlled Development.* St John's: Institute of Social and Economic Research.

Humphries, Vic. 1982. "Tragedy at Hibernia." *Oil Week*, February 22.

Hyams, Paul. 2003. *Rancour and Reconciliation in Medieval England.* Ithica: Cornell University Press.

Hynes, Ron. 1996. "Atlantic Blue." *Sold For a Song*. Peermusic Canada.

Ingram, David. 2010. *Habermas: Introduction and Analysis.* Ithaca: Cornell University Press.

Issacharoff, Samuel, and R.H. Klonoff. 2009. "The Public Value of Settlement." *Fordham Law Review* 78: 1177–27.

Jervis, Rick. 2011. "Oil Spill Victim Feels Forgotten." USA *Today*, April 13.

Jeudwire, J.W. 1917. *Tort, Crime and Police in Medieval Britain.* London: Williams and Nortgate.

Johnson, Paul. 2002. "Twenty Years On, Questions Remain." *Telegram* February 22.

Jones, Bryony. 2011. "BP Profit Hit $5.3 Billion, Year on from Gulf Spill." CNN on line, July 26.

Jones, Christopher. 2010. "Damage Caused by Transocean Deepwater Horizon Explosion — A Brother's Statement." Testimony Before the Senate Committee on Judiciary, United States Senate. June 8.

____. 2011. Interview, July 15.

Joyce, Randolph. 1982. "The Cruel Sea." *Macleans*, March 1.

Kant, Immanuel. 1998. *Groundwork of the Metaphysics of Morals.* (Mary J. Gregor, ed. and trans.). Cambridge: Cambridge University Press.

Kaplan, E. Ann. 2005. *Trauma Culture: The Politics of Terror and Loss in Media and Literature.* London: Rutgers University Press.

Kesselring, K.J. 2003. *Mercy and Authority in the Tudor State.* Cambridge: Cambridge University Press.

Kingsley, Charles. 1849. "The Three Fishers." <http://www.bartleby.com/246/572.html>.

Klar, Lewis. 2003. *Tort Law.* Thompson and Carswell: Toronto.

Kraus, Charles. 1983. Letter from Charles Kraus to Francis O'Dea, October 11

____. 1983b. Letter from Charles Kraus to Wells, O'Dea, Halley, Earle, Shortall and Burke, October 11.

____. 1983c. Letter from Charles Kraus to Noreen O'Neill, October 18.

____. 1983d. Letter from Charles Kraus to Noreen O'Neill, October 27.

____. 1983e. Letter from Charles Kraus to Raymond Halley, November 8.

____. 1983f. Telex from Charles Kraus to Leo Barry, December 7.

Kreindler and Kreindler. 1983. Letter from Paul Edelman to Scotty Morrison. November 29.

____. 1984. Letter from Paul Edelman to Scotty Morrison. June 11.

____. 1982. Letter from Paul Edelman to Lou Paisley. December 1.

LaCapra, Dominick. 2004. *History in Transit: Experience, Identity, Critical Theory.* Ithaca: Cornell University Press.

____. 2001. *Writing History, Writing Trauma.* Baltimore: Johns Hopkins University Press.

Lalonde, Marc. 1982. Commons Debates. *Hansard*, February 16: 15044

Laxer, Jim. 2008. *Oil*. Anansi: Toronto.

Leader Post. 1982a. "Political Opponents Called Traitors." April 5.

____. 1982b. "Newfoundlanders Get Day Off to Protest Ottawa Oil Moves." May 20.

____. 1982c. "Local Lawyers Are Upset." March 3.

Leger, Dan. 1982. "United States Launches Inquiry into Sinking of Ocean Ranger." *Evening Telegram*, April 27.

LeGoff, Jacques. 1990. *Your Money or Your Life: Economy and Religion in the Middle Ages*. (Trans. P Ranum.) Boston: MIT Press.

Lethbridge Herald. 1983. "Spill Report Stopped by Rig Owners." March 11.

Lindahl, Carl (ed.). 2001. *Perspectives on the Jack Tales and Other North American Marchen*. Bloomington: Folklore Institute, Indiana University.

Lintott, Andrew. 1999. *Violence in Republican Rome*. Oxford: Oxford University Press.

Long, Senator Russell. 1982. "Sneak Move Will Bar Claims by Rig Workers." *Globe and Mail*, December 14.

Lovelace, Martin. 2001. "Jack and His Masters: Real Worlds and Tale Worlds in Newfoundland folktales." *Journal of Folklore Research* 38, 1/2. Special Double Issue: Perspectives on the Jack Tales and Other North American Märchen (Jan.–Aug.): 149–70.

____. 1997. "Motifing 'Folktales of Newfoundland.'" *Journal of Folklore Research* 34, 3 (Sep.–Dec.): 227–31.

MacDowell, Douglas. 1978. *The Law in Classical Athens*. Ithaca: Cornell University Press.

Macpherson, C.B. 2011 [1962]. *The Political Theory of Possessive Individualism: Hobbes to Locke*. Oxford: Oxford University Press.

Marland, Alex. 2010. "Masters of Our Own Destiny: The Nationalist Evolution of Newfoundland Premier Danny Williams." *International Journal of Canadian Studies* 42.

Marshall, William. 1985. "Ministerial Statement." NL Mines and Energy. July 10.

Marx, Karl. 1996. "The Eighteenth Brumaire." In Eagleton and Milne (eds.), *Marxist Literary Theory*. Oxford: Blackwell Publishers.

Marx and Engels. 1996. "The German Ideology." In Eagleton and Milne (eds.), *Marxist Literary Theory*. Oxford: Blackwell Publishers.

McNulty, Sheila. 2011. "Transocean Awards Bonuses 'For Safety' in 2010." *Financial Times* April 2.

McQuaid, John. 1993. "Collins Resigns Federal Judgeship; Resignation Letter is Given to Clinton." *Times-Picayune*, August 7.

Medicine Hat News. 1983. "Safety Offshore." February 11.

Mercer, Rick. 2005. "Memorial Day in Newfoundland." Rick Mercer's Blog. rickmercer. <blogspot.com/2005/07/memorial-day-in-newfoundland.html> July 1.

Meyer, Elizabeth. 2004. *Legitimacy and Law in the Roman World: Tabulae in Roman Belief and Practice*. Cambridge: Cambridge University Press.

Michael, Lorraine. 2010. Interview, June 11.

Miller, William Ian. 2005. *Eye for an Eye*. Cambridge University Press. At <lib.myilibrary.

com.ezproxy.library.dal.ca?ID=43189>.
Mobil Oil. 1982. *Dimensions* 4, 1 (March 1).
Montreal Gazette. 1985. "Newfoundland's Politicians Don't Pull Their Punches." December 25.
___. 1984. "Rig Victim's Families Feel Angry, Betrayed." August 17.
___. 1983. "Judge Dismisses U.S. Ranger Suits." June 25.
___. 1982. "Brian Peckford 'Wages War' for His Newfoundland." April 3.
___. 1981. "Lalonde 'Sad' at Nfld Resignation." September 14.

Moore, Lisa. 2009. *February*. Toronto: Anansi Press.
Moore, Oliver, and Tavia Grant. 2011. "Welcome to St John's: The Hottest Job Market in the Country." *Globe and Mail*, May 6.
Moores, Douglas. 2011. Interview, July 22.
Morrison, Ian "Scotty." 2010. Interview, May 25.
Morrison v. Ocean Drilling and Exploration Company et al. 1983. United States District Court, January 13.
Musselwhite, Benton Sr. 2010. Interview, June 20.
___. 2010b. "Detailed Biographical Sketch of Benton Musselwhite."
___. 2009. "For a More Democratic and Empowered UN." Benton Musselwhite at the United Nations <youtube.com/watch?v=vbK643qDyag>.
___. 1983. Letter to Judge Robert Collins. October 17.
Musselwhite, Benton Jr. no date. "About My Father and His Proven Record as a Great Trial Lawyer."
Musselwhite v. The State Bar of Texas. United States Court of Appeals, Fifth Circuit, Sept. 23, 1994, c.f. Musselwhite v. State Bar of Texas, US Court of Appeals, 5th Circuit, 42T.3d639
Musson, Anthony. 2001. *Medieval Law in Context: The Growth of Legal Consciousness from Magna Carta to the Peasants' Revolt*. Manchester: Manchester University Press.
Myer, Elizabeth. 2004. *Legitimacy and Law in the Roman World: Tabulae in Roman Belief and Practice*. Cambridge: University of Cambridge Press.
Narváez, Peter. 2002. "'I Think I Wrote a Folksong': Popularity and Regional Vernacular Anthems." *Journal of American Folklore* 115, 456, Folklore in Canada (Spring).
National. 1982a. CBC Archives. February 15.
___. 1982b. CBC Archives. February 22.
National Commission on the BP Deepwater Horizon Gulf Oil Spill and Offshore Drilling. 2011. Deep Water: The Gulf Oil Disaster and the Future of Offshore Drilling. <www.oilspillcommission.gov/final-report>.
National Transportation Safety Board. 1983a. "Capsizing and Sinking of the U.S. Mobile Offshore Drilling Unit OCEAN RANGER off the East Coast of Canada, 166 Nautical Miles East of St. John's, Newfoundland, on February 15, 1982." (NTSB-MAR-83-2).
___. 1983b. Recommendation Letter to Admiral Gracey, Commandant, U.S. Coast Guard. February 28.
Neville, Cynthia. 2010. *Land, Law and People in Medieval Scotland*. University of

Edinburgh Press: Edinburgh.

New York Times. 2011a. "Harsh Words for Regulators in Crisis Commission Report." January 27.

____. 2011b. "Financial Crisis Inquiry Commission." January 25.

____. 1991. "U.S. Judge Is Given Prison Sentence." September 17.

Newfoundland Herald. 1982a. "An Election Charged with Emotion." March 22.

____. 1982b. "Peckford's Well-Oiled Campaign." March 27.

____. 1982c. "The Person Behind the News Releases: Susan Sherk." March 20.

Newhook, Cle. 2008. Interview, December 11.

____. Undated. "Origin and Development of the Ocean Ranger Families Foundation." Unpublished.

Nichols, David. 1985. *The Domestic Life of a Medieval City: Women, Children, and the Family in Fourteenth-Century Ghent*. University of Nebraska Press: London.

Nishman, Robert F. 1991. "Through the Portlights of the Ocean Ranger: Federalism, Energy, and the American Development of the Canadian Eastern Offshore, 1955–1985." Master's thesis, Queen's University.

NLPC Party. 1982. "The Challenge 1982." Newfoundland and Labrador PC campaign leaflet.

O'Dea, Francis. 1983. Letter to Noreen O'Neill, August 18.

____. 1983b. Letter to Noreen O'Neill, February 3.

____. 1983c. Letter to Noreen O'Neill, May 10.

____. 1983d. Letter to Noreen O'Neill, July 6.

____. 1983e. Letter to Noreen O'Neill, July 27.

____. 1983f. Letter to Noreen O'Neill, December 9.

O'Neill, Brian. 1988. "The Myth and Reality of Offshore Oil and Gas Development: A Critical Inquiry into the Political Economy of Hydrocarbon Resource Development in the Offshore Regions of Newfoundland and Nova Scotia." Masters thesis, Saint Mary's University: Halifax.

____. 1987. "The Sinking of the Ocean Ranger, 1982: The Politics of a Resource Tragedy." *People, Resources and Power*. Fredericton: Gorsebrook Research Institute: 153–61.

____. 1985. "February 15, 1982: Where to Point the Finger." *New Maritimes* 2, 5 (February).

____. no date. "Safety in the Offshore." St John's: The Regional Centre for the Study of Contemporary Social Issues.

O'Neill, Noreen. 2011. Thank you on behalf of the Ocean Ranger families to the Gonzaga High School's memorial services' organizers and attendees.

____. 2010. Interview, June 14.

Obama, Barack. 2010. "Obama's Remarks on the Gulf Oil Spill." May 27. Transcript from the Council on Foreign Relations <cfr.org/united-states/obamas-remarks-gulf-oil-spill-may-2010/p22235>.

Ocean Ranger Families Foundation. 1985. Press Release on "Summary of Action Taken by the Government of Canada in relation to the Recommendations of the Royal Commission on the Ocean Ranger Marine Disaster." April 18.

ODECO *v Hickey Estate et al.* 1985. Newfoundland Supreme Court Trial Division. June 18. Lang.

ODECO *v Tilley's Estate et al.* 1985. Newfoundland Supreme Court Trial Division. June 18. Noel J.

Offshore. 1983. "Follow Up: ODECO Filed a $165 Million Claim Against Nine Companies." May: 8.

___. 1982. "Insurers Expect Rates to Remain Stable." April.

Oil and Gas Journal. 1982a. "Ocean Ranger Vanishes off Canada." February 22.

___. 1982b. "Suits Seek Damages in Ocean Ranger Sinking." March 29.

___. 1982c. "Drilling-Production: Ocean Drilling and Exploration Co. Received Insurance Payments Totaling $86.5 Million." August 9.

Oilweek. 1985."Memorial Site for Ocean Ranger." July 29.

___. 1983. April 4.

___. 1982. "Ocean Ranger Sinking Shocks Canada." February 22.

Olson, Trisha. 2006/2007. "The Medieval Blood Sanction and the Divine Beneficence of Pain: 1100–1450." *Journal of Law and Religion* 22, 1: 63–129.

Orlando Sentinel. 1985. "Oil Rig Disaster." April 15.

Ottawa Citizen. 1983. "Grit Leader Urges Criminal Charges." February 10.

Parsons-Walker, Cynthia. 2010. Interview, June 12.

Payne, Lana. 2011. "Improving SAR Response Times: It's a Matter of Life and Death." NLFL President presentation to the Standing Committee on National Defence — Delta Hotel, St. John's, NL. February 2.

Pearce, Frank. 1993. "Corporate Rationality as Corporate Crime." *Studies in Political Economy* 40: 135–62.

Peckford, Brian. 1982a. "A Mandate to Negotiate — Jobs and a Future." Address by the Honourable A. Brian Peckford, March 15.

___. 1982a. Press statement, March 22.

Pervukhin, Anna. 2005. "Deodands: A Study in the Creation of Common Law Rules." *American Journal of Legal History* 47, 3 (July).

Piercy Estate v General Bakeries Ltd. 1986. Supreme Court of Newfoundland Trial Division. Chief Justice Hickman.

Posner, Eric, and Cass Sunstein. 2005. "Dollars and Death." *University of Chicago Law Review* 72, 2: 537–98.

Postmedia News. 2004. "PM Still Facing Pressure to Honour Commitment to Newfoundland Premier." November 5.

Prince Rupert News. 1983. "Ocean Ranger Settlement $20m." October 13.

Pross, A. Paul, Innis M. Christie, and John A. Yogis. 1990. *Commissions of Inquiry.* Toronto: Carswell.

Quinn, Mark. 2010. "N.L. Chopper Crash Widow Assails Companies." CBC news. February 10.

Robinson, Terry. 1992. *The Offshore.* St John's: Jesperson Press.

Rousseau, Jean-Jacques. 1978. *On the Social Contract.* (Masters, trans.). New York: St. Martin's Press.

RC (Royal Commission on the Ocean Ranger Marine Disaster). 1984. "Report One:

The Loss of the Semisubmersible Drill Rig Ocean Ranger and its Crew." Canadian Government Publishing Centre, Publication Number Z1-1982/1-1E, ISBN 0-660-1 1682-0, pp. iii-iv, August.

____. 1985. "Report Two: Safety Offshore Eastern Canada."

Rupke, Jorg. 1992. "You Shall Not Kill. Hierarchies of Norms in Ancient Rome." *Numen* XXXIX, 1 (June): 58–79.

Ryan, Patricia. 2010. Interviews May 11, 22 and June 8.

Safina, Carl. 2011. *A Sea in Flames*. New York: Crown Publishing.

Schuck, Peter H. 2004. *Agent Orange on Trial: Mass Toxic Disasters in the Courts*. Cambridge, MA: Belnap Press.

Schwartz, John. 2011. "Senate May Alter Law for Widows from Rig Blast." *New York Times*, June 7.

Sider, Gerald, and Gavin Smith. 1997. *Between History and Histories: The Making of Silences and Commemorations*. Toronto: University of Toronto Press.

Smith, Dorothy. 1999. *Writing the Social: Critique, Theory and Investigations*. Toronto: University of Toronto Press.

____. 1990. *The Conceptual Practices of Power: a Feminist Sociology of Knowledge*. Toronto: University of Toronto Press.

Spagnoletti, Frank. 2010. Interview, July 21.

Statistics Canada. 2011. <www.statcan.gc.ca/start-debut-eng.html>.

Telegram. 2010. "Offshore Oil Industry Needs Strong Regulator, Union Lawyer Tells Inquiry." September 9.

____. 2007. "Runaround: Ocean Ranger Film Shut Down, No Rig to Shoot On." June 17.

Telegraph Journal. 1983. "Out-Of-Court Settlement Said Close in Ocean Ranger Sinking Disaster." December 23.

____. 1982. "Effect of Ocean Ranger Disaster Now Hitting." July 23.

Texas Lawyer. 2011. "Dole Settles Pesticide Claims with 5,000 Former Banana Plantation Workers." June 15.

Thomas, Gerald, and J.D.A. Widdowson (ed.). 1991. *Studies in Newfoundland Folklore: Community and Process*. Toronto: Breakwater Books.

Tiller, Greg. 2009. *Random Thoughts: Memories & Writings of Greg Tiller*. (Compiled by Steve Porter.) D.C. Publishing. Reprinted in *Evening Telegram* March 6, 1983.

Toronto Star. 1983. "Families of Oil Rig Dead Awarded up to $900,000." December 29.

Tucker, Eric. 1990. *Administering Danger in the Workplace: The Law and Politics of Occupational Health and Safety Regulation in Ontario 1850–1914*. Toronto: University of Toronto Press.

Tucker, Norman A. 1983. "One Underwriter's View on Safety Offshore." Proceedings from the Symposium on the Safety of Life Offshore International Association of Drilling Contractors and Scripps Institution of Oceanography, June 8–10. San Diego: University of California

U.S. Coast Guard. 1983. "Marine Casualty Report Mobile Offshore Drilling Unit (MODU) OCEAN RANGER, O.N. 615641, Capsizing and Sinking in the Atlantic Ocean on 15 February 1982 with multiple loss of life." May. Washington: U.S. Department of Transportation.

____. 1978. "Marine Board Casualty Report OCEAN EXPRESS (Drilling Unit); Capsizing and Sinking in the Gulf of Mexico on 15 April 1976 with loss of life." June. Washington: U.S. Department of Transportation.

U.S. Congress, House Committee on Merchant Marine and Fisheries. 1982. Hearings on Ocean Ranger collapse. Washington.

Van Der Kolk, Bessel, and Onno Van Der Hart. 1995. "The Intrusive Past: The Flexibility of Memory and the Engraving of Trauma." In Cathy Caruth (ed.), *Trauma: Explorations in Memory*. Baltimore: John's Hopkins University Press.

Verberg, Norreen, and Chris Davis. 2011. "Counter-Memory Activism in the Aftermath of Tragedy: A Case Study of the Westray Families Group." *Canadian Review of Sociology* 48, 1: 23–45.

Victoria Advocate. 1989. "Attorneys Targeted in Texas Probe." January 9.

Voyer, Roger. 1983. *Offshore Oil: Opportunities for Industrial Development and Job Creation*. Ottawa: Canadian Institute for Economic Policy.

Wadden, Marie. 2010. Interview, June 15.

____. 1983. Unpublished notebook.

Wagner, Diane. 1986. "New Elite Plaintiff's Bar." ABA *Journal* 72 2: 44–49.

Wagner, Ray. 2010. Interview, June 22.

Wagner-Pacifini, Robin. 2010. "Theorizing the Restlessness of Events." *AJS* 115, 5 (March): 1351–86.

Warnke, Georgia.1995. "Communicative Rationality and Cultural Values." In White (ed.), *The Cambridge Companion to Habermas*. Cambridge: Cambridge University Press.

Warren, Ted. 1983. "Ocean Ranger Families: Reports of Claim Settlement Premature?" *Newfoundland Herald*, October 29.

Weber, Max. 2002. *The Protestant Ethic and the Spirit of Capitalism*. (Kalberg trans.) Cary: Roxbury Publishing.

Weinstein, Jack B. 2009. "Comments on Owen M. Fiss, 'Against Settlement' (1984)." *Fordham Law Review* 78, 3.

Werner, Erica. 2010. "Obama Blasts Big Oil, Regulators: U.S. President Pledges to End 'Cozy Relationship' Between Companies, Federal Government." AP/*Chronicle Herald* May 15.

White, G. Edward. 1980. *Tort Law In American: An Intellectual History*. Oxford: Oxford University Press.

White, Hayden. 2010. *The Fiction of Narrative: Essays on History, Literature and Theory 1957–2007*. Baltimore: Johns Hopkins University Press.

Williams, Danny. 2010. Speech. *National Post* transcript November 25.

Wilson v ODECO *et al*. 1985. Affadavit in the Supreme Court of Newfoundland, April 1.

Witt, John Fabian. 2004. *Accidental Republic: Crippled Workingmen, Destitute Widows and the Re-Making of American Law*. Boston: Harvard.

Wood, Allen. 2004. *Karl Marx*. New York: Routledge.

Wylie, Herb. 2010. "February Is the Cruelest Month: Neoliberalism and the Economy of Mourning in Lisa Moore's February." *Newfoundland and Labrador Studies* 24, 1: 1–11.

Zelizer, Viviana. 1994. *Pricing the Priceless Child: The Changing Social Value of Children*. Princeton University Press.
Zierler, Amy. 1982a. "Uncertainty Will Delay Hibernia." *Financial Post,* February 20.
___. 1982b. "To Sinking a Titanic." CBC radio documentary, October 24.
___. 1983a. "Ocean Ranger Negotiations Lumber On." *Financial Post,* October, 22.
___. 1983b. "Offshore Ownership Battles Continue as Mobil Pulls Oil Rigs into Harbour." *Financial Post,* February 26.
Zierler, Amy, and Roger Bill. 1982. "To Sinking a Titanic." CBC radio documentary (transcript). October 24.

INDEX